Interdisciplinary Applied Mathematics

Volume 63

Problems in engineering, computational science, and the physical and biological sciences are using increasingly sophisticated mathematical techniques. Thus, the bridge between the mathematical sciences and other disciplines is heavily traveled. The correspondingly increased dialog between the disciplines has led to the establishment of the series: Interdisciplinary Applied Mathematics.

The purpose of this series is to meet the current and future needs for the interaction between various science and technology areas on the one hand and mathematics on the other. This is done, firstly, by encouraging the ways that mathematics may be applied in traditional areas, as well as point towards new and innovative areas of applications; and secondly, by encouraging other scientific disciplines to engage in a dialog with mathematicians outlining their problems to both access new methods as well as to suggest innovative developments within mathematics itself.

The series will consist of monographs and high-level texts from researchers working on the interplay between mathematics and other fields of science and technology.

Timothy G. Myers

Theoretical and Practical Stefan Problems

Springer

Timothy G. Myers
Campus de Bellatera, Edifici C
Centre de Recerca Matemàtica
Bellaterra, Barcelona, Spain

ISSN 0939-6047 ISSN 2196-9973 (electronic)
Interdisciplinary Applied Mathematics
ISBN 978-3-032-04825-7 ISBN 978-3-032-04826-4 (eBook)
https://doi.org/10.1007/978-3-032-04826-4

Mathematics Subject Classification: 80A22, 80-01

This Springer imprint is published by the registered company Springer Nature Switzerland AG
The registered company address is: Gewerbestrasse 11, 6330 Cham, Switzerland

Many colleagues, students, and post-docs have been involved in the research incorporated into this book. Particular mention goes to Francesc Font (U. Politècnica de Catalunya), Matt Hennessy (U. Bristol), Vincent Cregan (Cork Inst. Tech.), Michelle MacDevette (gone off the rails! Well, into financial mathematics), Marc Calvo (U. Politècnica de Catalunya), Helena Ribera, Gift Muchatibaya (U. Zimbabwe), Jean Charpin (also lost to finance), and presumably many others whom I have forgotten to acknowledge. Also thanks to Lucy Auton (CRM) who made the figures so much prettier than my original versions!

Prof. Sarah Mitchell of the University of Limerick deserves a special mention. My most constant collaborator, she was involved in much of the work described here. She was also my numerical champion. I'm sorry she is no longer around to read this (and, no doubt, point out errors).

Most importantly I dedicate this book to my children Anna and Tomás. Remember that in research as in life it is important to be honest. It may not make you popular but at least you will sleep well.

"The things of this world cannot be made known without a knowledge of mathematics"
Roger Bacon 1214–1292

Preface

Stefan problems, a sub-class of moving boundary problems, have been the subject of investigation for over 150 years. The original problem, studied by Jožef Stefan and published in 1889, concerned the formation of sea-ice. This has since been extended to many forms of phase change, including melting, sublimation, vaporisation, and deposition, and beyond this, there exists a wide variety of analogous problems. Drawing together the different forms and applications, the loose term, *Stefan problems*, forms an important and wide ranging field of research.

My own introduction to this topic came over 30 years ago, when I was contracted to model water flow over an accreting ice surface for a consortium of aerospace companies. At the time the prevalent method to track the ice growth was a one-dimensional equilibrium, isothermal, energy balance known as the Messinger model. I pointed out that this method was very dated and could easily be improved—they asked for a demonstration. Some months later at a consortium meeting I presented a comparison of my model with an actual ice growth. I predicted a change in the formation process at around 2.6 mm, their measurement was between 2 and 3 mm. The room burst into applause. Afterwards I asked one of them why they had clapped, he explained it was because no-one thought a mathematician could do something so practical. I wrote up the model, which is now used in a number of icing codes, and referred to it as an extension to the Messinger model (see Chap. 3). The reason for implying that it was an extension was to not frighten off the practitioners but to imply that I was really just extending what they already use rather than introducing some obscure mathematical theory. In fact it is not an extension, it is a very different form of model derived with strict bounds and limitations. In one, unphysical limit, it may be reduced to the Messinger case. Every now and then it is referred to as the Myers model, but much more commonly as the Extended Messinger Model. We will see later that the developer of a model seldom gets credit, so I should not be too bitter that my first foray into the world of Stefan problems enhanced Messinger's name and not my own.

My *modus operandi* is to develop a physical model and then look for reductions which permit some analytical progress or at least simplify numerical calculations. In this book I stick with this approach and use my experience in the field to motivate a range of mathematical models. Since there already exists a number of excellent books on standard models I describe some of the recent advances in the theory of Stefan problems, with a focus on physical applications and situations where analytical progress is possible. The initial chapters provide a, hopefully, gentle introduction into the theory and analysis of classical one- dimensional Stefan problems. The one-dimensional restriction is not as restrictive as it sounds, since problems can easily be broken up into different compartments. For example, in the case of aircraft icing the one-dimensional ice growth is generally coupled to three-dimensional water flow. It is also, obviously, the prerequisite for more complex models. Chapter 3 deals with phase change in the presence of a flowing liquid. Due to the complexity of adding flow, this is limited to thin film or almost unidirectional flows, with applications to solidification in pipes, contact melting, and atmospheric icing. In Chap. 4 we move away from the classical Stefan problem to develop a model where material properties or the phase change temperature may vary through the process. This leads to a more complex form of Stefan condition, including kinetic energy and sensible heat. The issue of energy conservation in the one-phase reduction is then analysed subject to the new form of Stefan condition. Chapter 5 deals with situations where the phase change temperature is a variable, with applications in the solidification of supercooled melts, nanocrystal growth (a mass diffusion problem), and nanoparticle and nanowire melting. The final chapter focuses on phase change at the nanoscale, where heat flow no longer follows Fourier's law (which may be thought of as an average description of the frequent collision between phonons). At the nanoscale heat flow may be described by the infrequent collision of phonons and so must be treated as ballistic rather than diffusive. This removes the unrealistic infinite speed of heat propagation implicit to Fourier's law but introduces a new level of mathematical complexity.

The book is aimed at anyone interested in Stefan or moving boundary problems. The initial chapters follow a relatively standard format, with an almost classical formulation and are suitable for final year or graduate level students. Later chapters deal with more complex topics, going beyond the classical Stefan problem and are aimed more at the graduate level. Late in the writing process I added exercises, I hope there are not too many mistakes or difficulties, just send me an email!

In fact there remains many possibilities for applying and extending the topics of the later chapters to new research problems. So read the book, digest it, and take the work to even more fascinating realms.

Bellaterra, Barcelona, Spain
August, 2025

Timothy G. Myers

Declarations

Competing Interests The author has no competing interests to declare that are relevant to the content of this manuscript.

Contents

Chapter 1
Introduction

1.1 Background

The classical Stefan problem is a form of moving boundary problem that originally described the liquid to solid phase change process. It is named after Jožef Stefan who published a series of four papers on heat conduction and phase change in 1889 [145–148]. One of these, concerning the formation of ice from the water of the polar seas, is the basis for this physically important field of study. However, it is worth pointing out that in 1831 Lamé and Clapeyron studied an equivalent problem when investigating the formation of the Earth's crust. They imposed a fixed temperature boundary condition, assumed the molten material was at the phase change temperature and determined that the crust thickness grew with the square root of time. Stefan found the same square root dependence but he also determined the constant of proportionality in the growth rate: a quantity which had eluded Lamé and Clapeyron. Even with this advance Stefan was not the first, the solution had already been presented by Neumann in a series of lectures given in the early 1860s. Neumann found a similarity solution for the solidification of a semi-infinite media, initially above the freezing temperature, when at $t = 0$ the boundary $x = 0$ is set to a constant temperature below the freezing value. Since the lecture notes were not published for another 40 years, perhaps Stefan can be forgiven for not crediting Neumann. Problems where the temperature of one of the regions is set to its phase change value are now termed one-phase Stefan problems, Neumann solved the two-phase problem (where the temperature varies in both regions). An interesting history of the Stefan problem may be found in Rubinstein's book [134].

To summarise, the system formulated by Lamé and Clapeyron and first solved by Neumann is now termed the Stefan problem. Research (and life) can be like this.

In this book, solely to reduce the number of physical explanations and to avoid confusing the notation, we will work primarily with the most common interpretation of the Stefan problem where phase change is driven by temperature gradients. However, there exist many other relevant physical situations, and Stefan problems

T. G. Myers, *Theoretical and Practical Stefan Problems*, Interdisciplinary Applied Mathematics 63, https://doi.org/10.1007/978-3-032-04826-4_1

fall into the wider category of *free boundary problems* where a PDE system must be solved over an unknown domain. In the specific case where the domain evolves, these may be termed *moving boundary problems*. In Crank [36], a wide range of models and mathematical techniques related to moving boundary problems in the context of mass diffusion are presented. These are motivated by situations where molecules in a gas or liquid are somehow immobilised, for example, by attaching to adsorption sites. Other situations where free boundaries occur include flow in porous media, mathematical finance, shock waves in gas dynamics, contaminant removal from liquids, crack growth in solids and polymer, and grain swelling [11, 36, 118, 123]. Since these are mathematically analogous, the analysis and results are equivalent to those presented in the following chapters.

There are a number of textbooks dealing with different aspects of phase change. Alexiades and Solomon [4] work at the interface between mathematics and engineering, presenting techniques to deal with physically realistic situations. Poots [129] takes a purely pragmatic approach in a book devoted to ice and snow accretion on structures. Davis [40] is motivated primarily by problems in materials science. Gupta [59] focusses on the mathematical properties and concepts involved in the study of moving boundary problems. Hill [64] reviews past research and presents approximate solution methods to the one-dimensional Stefan problem, with a strong emphasis on polynomial approximations. The two texts by Carslaw and Jaeger [29] and Crank [35] cannot be overlooked despite the fact they deal primarily with diffusion over a fixed domain, the first concerning the flow of heat and the second of mass. However, they are both fonts of essential information and both contain a whole chapter on moving boundary problems.

The present text is most closely aligned to that of [4] in the sense that, while primarily mathematical in nature, all problems studied are motivated by physical situations. For example, in Chap. 3, we consider phase change with a flowing liquid in the context of microvalves and ice accretion on aircraft. In Chap. 5, we consider the solidification of a supercooled liquid, the melting of nanoparticles, and the growth of nanocrystals. Finally, in Chap. 6, we investigate phase change when the heat flow no longer follows Fourier's law, as is the case at the nanoscale or extremely short heating events (e.g. ultra-fast laser heating [81, 131]). The topics investigated have been published in mathematics, physics, engineering, and chemistry journals over the past 20 years and are all current active research areas.

1.2 Governing Equations

Temperature-driven phase change involves the energy-driven transformation of mass from one form to another. The mathematical description of the process requires specifying heat equations in either phase. The extent of each phase is unknown *a priori* and changes with time. To determine the position of the moving boundary, a further equation is required. For thermal problems, this is an energy balance, the

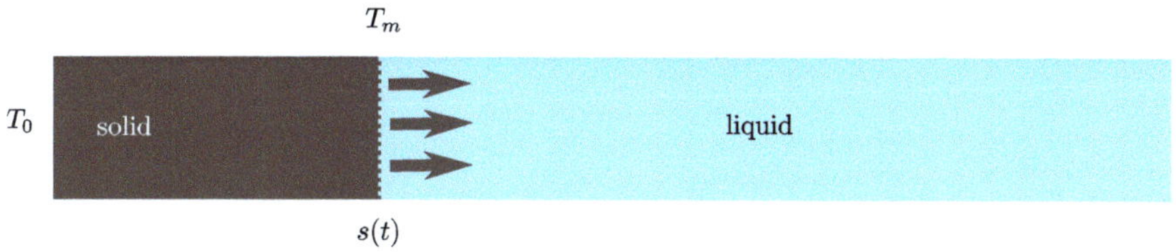

Fig. 1.1 Typical one-dimensional phase change configuration. A liquid, initially occupying the domain $x \geq 0$, is cooled at $x = 0$ below the phase change temperature, T_m, such that solidification occurs. The position of the moving solidification interface is denoted $x = s(t)$, where $s(0) = 0$

Stefan condition, which relates the velocity of the boundary to the energy jump across the interface.

The heat equations and Stefan condition may be derived in a number of different ways. To ease the reader into the subject here, we will simply state the basic governing equations, which are subject to a number of restrictions. In Chap. 4, we will derive a more general system, using conservation laws, to provide a set of equations suitable for moving fluid layers, variable phase change temperatures, curved surfaces, and a pressure jump at the interface.

We begin by stating a number of common assumptions (many of which we will neglect in the course of this book):

- Gravity and viscous effects are negligible.
- The thermophysical properties, such as specific heat and thermal conductivity, are constant within each phase (often they are set to the same value in both phases).
- The phase change temperature is constant.
- The density, ρ, is constant and the same in both phases.
- The interface between the two phases is sharp (i.e. of zero width).

A more complete list of standard assumptions, together with a discussion on their realism, may be found in Alexiades and Solomon [4, Table 1.1]. In particular, they state that the constant density assumption is possibly the most unreasonable, as anyone who has had a burst water pipe can verify.

Now we focus on the one-dimensional Cartesian system. A typical configuration is shown in Fig. 1.1. A semi-infinite liquid is solidified by imposing a temperature T_0 at the left-hand boundary, $x = 0$, where T_0 is below the phase change temperature, $T_0 < T_m$. The interface between solid and liquid regions is denoted $x = s(t)$ and moves with velocity $s_t > 0$.

The heat equations describing a system such as that of Fig. 1.1 are a statement of energy conservation within either phase. For a solid-liquid system, the one-dimensional heat equations are

$$\rho_s c_s \frac{\partial T}{\partial t} = k_s \frac{\partial^2 T}{\partial x^2}, \qquad 0 < x < s(t), \tag{1.1a}$$

$$\rho_l c_l \frac{\partial \theta}{\partial t} = k_l \frac{\partial^2 \theta}{\partial x^2}, \qquad s(t) < x, \tag{1.1b}$$

where T is the solid temperature, θ the liquid temperature, ρ the density, c the specific heat capacity, and k the thermal conductivity. The subscripts s and l denote solid and liquid, although we could equally well be thinking of evaporation or condensation (hence liquid-gas), sublimation, or deposition (solid-gas or gas-solid). These equations hold over adjacent domains separated by the boundary $x = s(t)$. In the one-dimensional case, this position is an unknown, time-dependent point. It is determined via an energy balance across the interface, the *Stefan condition*:

$$\rho_s L_m \frac{ds}{dt} = k_s \frac{\partial T}{\partial x}\bigg|_{x=s} - k_l \frac{\partial \theta}{\partial x}\bigg|_{x=s}, \tag{1.2}$$

where the derivatives are calculated at the interface $x = s$ (strictly we may prefer to think of $x = s^+, x = s^-$, *i.e.* immediately adjacent to the interface). The right-hand side denotes the difference between the energy conducted through the solid and the liquid. In the absence of a phase change, $s_t = 0$, these terms are equal, and we are left with a standard continuity of heat flux boundary condition. Thinking in terms of a liquid-solid transition, the molecular structure of a solid and liquid is very different, and liquids have the higher internal energy (liquid molecules move easily and have a higher kinetic energy than those of a solid). Consequently, energy must be supplied to a solid to melt it; conversely freezing leads to a release of energy. This difference in internal energy between phases means that when a phase change occurs, there is a change in the internal energy; this jump is termed the *latent heat* (specifically latent heat is measured at a constant pressure). Generally, the latent heat is the same whether we move from liquid to solid or solid to liquid. We will denote the latent heat by L_m (which comes from latent heat of melting; however, subscripts f for fusion and v for vaporization are also common). Note, if the difference in densities between the two phases is taken into account, then it is the solid density appearing on the left-hand side of Eq. (1.2) (this will be demonstrated in Chap. 4).

Equation (1.2) states that the rate at which energy is released due to the change of phase is balanced by the heat removal through the solid and liquid regions.

1.2.1 Boundary and Initial Conditions

The governing equations consist of two second-order parabolic partial differential equations, describing the temperature, and an energy balance at the interface. This apparently requires initial conditions for the two temperatures and the interface position as well as two boundary conditions per heat equation. We begin by examining the boundary conditions.

At the interface $x = s(t)$, the temperature takes the phase change value which, in standard problems, is constant $T(s, t) = \theta(s, t) = T_m$ (in later chapters, we will permit variation); this specifies two of the required boundary conditions. The remaining conditions, away from the melt interface, may take the form:

1. Specified temperature (Dirichlet condition)

$$T = T_0 . \tag{1.3}$$

2. Specified heat flux (Neumann condition)

$$k\frac{\partial T}{\partial x} = -q . \tag{1.4}$$

3. Heat flux proportional to a temperature difference (Newton cooling or Robin condition)

$$k\frac{\partial T}{\partial x} = \mathscr{H}(T_a - T) . \tag{1.5}$$

The values, T_0, q, T_a, may vary with time (and with space when we move away from one-dimensional problems). Identifying $-kT_x$ as the flux of heat, the second and third conditions represent heat flow caused by a specified energy flux or by a temperature jump across the boundary. The heat transfer coefficient $\mathscr{H}$ is usually measured experimentally and depends on quantities such as the material being studied, the surface roughness, and, in the case of a fluid, the form of the flow. The third condition, Eq. (1.5), is clearly the most general form. In the limit of infinite heat transfer, $\mathscr{H} \to \infty$, the Dirichlet condition, Eq. (1.3), is retrieved. If $T_a \gg T$, then the Neumann condition, Eq. (1.4), is retrieved by defining $\mathscr{H}T_a = -q$. This might suggest applying the Robin condition in all situations and then other forms may be obtained as limiting cases. Unfortunately, it leads to the most complex expressions for the temperature, which are often not readily usable. Despite the fact that in reality it can be very difficult (one could easily insert the words virtually impossible here) to impose a fixed temperature throughout a process, the vast majority of mathematical work focusses on the first condition. Often in this book, we will use this boundary condition as a starting point, since solutions are readily available.

Any reader interested in the application of realistic boundary conditions may wish to peek now at Sect. 5.4.1, where it is shown that infinite $\mathscr{H}$ is not possible (an infinite value would lead to the material vaporising). Further, even with the maximum possible physical value, in a study of nanowire melting [49], it is shown that choosing the largest physically possible heat transfer coefficient leads to significantly different results when compared to those obtained by applying a Dirichlet condition. So, while the Dirichlet condition is ubiquitous and convenient, it should be borne in mind that this infinite $\mathscr{H}$ limit is questionable (it could lead to explosions) and so should be treated with care in certain physical situations (such as those discussed in Chaps. 5 and 6).

If the domain is so large that it may be treated as semi-infinite, then in the far field, it is standard to assume the temperature approaches its initial value or that the derivative tends to zero. This also indicates that "large" could be interpreted in the sense that the process has been running for a sufficiently short time that the far-field temperature has not changed significantly from its initial value.

Despite specifying two heat equations, at $t = 0$ only one of the phases exists, so only a single initial condition, such as $\theta(x, 0) = T_i$, is required for the existing phase. However, try explaining this to a computer. For a numerical solution, a small time approximation may be required and then the program started at some early time $t = t_i$; this is discussed in more detail in Sect. 2.2.3 and 5.2.3. Finally, the Stefan condition requires an input value to pin down the phase change front. In the one-dimensional Cartesian case, this is usually defined as $s(0) = 0$.

Coupling the Stefan condition (1.2) to the heat equations (1.1), together with appropriate boundary conditions, defines the classical *Stefan problem*. However, the classical Stefan problem is not the focus of this text, so without any regrets, we immediately point out a number of limitations of this approach:

1. Phase change invariably results in a density jump across the interface and consequently one of the materials will have to move to make space. The resultant velocity and kinetic energy are often neglected. It may be argued that their contribution is negligible, but for the standard example of a fixed temperature boundary condition, where the boundary temperature differs from the melt temperature, $T(0, t) \neq T_m$, we will see in Chap. 2 that the initial velocity is infinite (suggesting that neither velocity nor kinetic energy is negligible!). This issue is discussed in Chap. 4.
2. While a one-dimensional, flat interface, located at $x = s(t)$, is nice to behold mathematically, it is not very common in the real world. In higher dimensions, the fluid may flow due to external forces, leading to solidification away from the initial domain. Solidification with flow is investigated in Chap. 3. If we consider a curved surface, then there will be surface tension between the phases. This has a corresponding surface energy, which will vary as the surface changes. This is discussed in Chaps. 4 and 5.
3. There are many physical situations where the phase change temperature varies during the process, for example, with supercooled liquids or at highly curved surfaces (where the pressure is high). This is dealt with in Chap. 5.
4. At very small scales, Fourier's law is known to break down, and so the standard heat equations no longer hold. This forms the focus of Chap. 6.

We may immediately partially rectify the second point by inferring a more general form for the Stefan problem:

$$\rho_s c_s \frac{\partial T}{\partial t} = k_s \nabla^2 T \,, \tag{1.6a}$$

$$\rho_l c_l \frac{\partial \theta}{\partial t} = k_l \nabla^2 \theta \,, \tag{1.6b}$$

$$\rho_s L_m \frac{dR}{dt} = (k_s \nabla T - k_l \nabla \theta)|_R \cdot \hat{\mathbf{n}}\,, \tag{1.6c}$$

where, anticipating treating spherical and cylindrical problems, we follow convention and denote the position of the interface by $R(t)$ and $\hat{\mathbf{n}}$ the unit normal. If there is any fluid flow, then Eq. (1.6b) may be replaced by

$$\rho_l c_l \left(\frac{\partial \theta}{\partial t} + \mathbf{u} \cdot \nabla \theta \right) = k_l \nabla^2 \theta\,, \tag{1.7}$$

where $\mathbf{u}$ is the fluid velocity vector. Fluid flow may be caused by the phase change process, when the densities are not equal, or by some external force such as gravity, pressure difference, or surface shear. In this text, we will assume that the solid does not deform and so set the solid velocity to zero, $\mathbf{u}_s = 0$ (even if there is solid motion, we can define our frame of reference as moving with the solid and retain the zero velocity condition). The above equations and variants are derived in many texts including [4, 40]. Crank [36] discusses generalisations such as variable thermophysical properties, space or time-dependent phase change temperature, and a heat source on the moving boundary or inside the material.

1.2.2 Mushy Regions

The majority of the cases examined in this text deal with phase change with a sharp interface. However, certain scenarios involve morphological instabilities; for example, when dendrites grow into the fluid layer, there is a region, of finite thickness, occupied by both liquid and solid phases. Solomon [143] suggests a scenario where ice, at zero degrees, is broken into small pieces and then placed into a pool of water, also at zero degrees. The resultant slush is neither pure liquid nor pure solid and cannot be described as a region with a sharp interface. Materials subject to internal heating, for example, through electrical resistance or radioactive decay, may heat up to their melt temperature. Subsequent heating will not result in a phase change until an amount of energy equal to the latent heat has been generated. A material at the phase change temperature, without actually changing phase or a mixture of solid and liquid at the phase change temperature, is termed *mushy* (as will be defined below) and the associated volume the *mushy region*.

A solid awaiting enough energy to turn to a liquid, a slush, or a mixture of dendrites and liquid appear physically distinct. One way to understand and relate these seemingly different states is through the *enthalpy*.

Enthalpy is a measure of the thermodynamic energy in a system. It is the sum of a system's internal energy and the product of the pressure with the volume (this latter term represents work done against an external pressure). Following this definition, we may write

$$\mathrm{H} = U + pV \tag{1.8}$$

where H, U denote the total enthalpy and internal energy. To convert to intensive properties, that is, ones that do not depend on the size, we divide by the mass to obtain the specific enthalpy:

$$\mathrm{h} = u + p/\rho\,. \tag{1.9}$$

In practice, it is not possible to directly measure enthalpy, since the exact components of the internal energy are generally unknown, instead the change in enthalpy is considered. The internal energy may be expressed in terms of the specific heat capacity, c, temperature, and pressure. In the often studied case where there is no change in pressure in the system, we define the change in specific enthalpy as

$$\Delta\mathrm{h} = c_p \Delta T \tag{1.10}$$

where c_p is the specific heat capacity measured at constant pressure. From this, the integrated form follows naturally:

$$\mathrm{h} = \begin{cases} \int_{T_m}^{T} c_s \, dT & \text{for } T < T_m \\ \int_{T_m}^{\theta} c_l \, dT + L_m & \text{for } \theta > T_m\,, \end{cases} \tag{1.11}$$

where c_s, c_l represent the specific heat capacity of the solid and liquid measured at constant pressure (of course with incompressible materials, constant pressure and constant volume are equivalent). Carrying out the integration in each region requires a different constant of integration. As discussed in Sect. 1.2, the latent heat represents the change in internal energy between the phases; choosing $\mathrm{h} = 0$ when the solid temperature is T_m, then the latent heat may be viewed as the jump in enthalpy between the two material states at the phase change temperature. (In Sect. 4.2.2, we discuss internal energy and heat flow in more detail.)

Returning to the concept of a mushy region, the solid region is identified as having $\mathrm{h} < 0$ the liquid as $\mathrm{h} > L_m$ and the mushy region as $0 < \mathrm{h} < L_m$. For constant specific heat in each phase, integration of (1.11), with appropriate conditions, gives

$$\mathrm{h} = \begin{cases} c_s(T - T_m) & \text{for } T < T_m \\ \in [0, L_m] & \text{for } T = T_m \\ c_l(\theta - T_m) + L_m & \text{for } \theta > T_m\,. \end{cases} \tag{1.12}$$

The value of the enthalpy is key to the *enthalpy method*, which is discussed in Sect. 2.4. In Sect. 1.2.2, we demonstrate an exact solution for the growth of a solid and mushy region, subject to various restrictions, developed in [144]. Following the work of Huppert and Worster [61] on binary alloys, Davis [40, §10.3] sets out a number of models for mushy regions where the mush temperature varies. This occurs due to concentration variation within the mush, which then alters the phase change temperature.

1.3 Nondimensionalisation

It is hard to impress upon the reader the importance of nondimensionalisation without showing practical examples. Indeed, many of the models presented in later chapters would be too difficult to solve without the benefits derived from the nondimensional form. Nondimensionalisation allows us to identify dominant and negligible terms, to reduce the parameter space of the problem and to work with variables of the order unity. Fowler [54] points out that physically realistic *as opposed to academically titillating* models are usually too complicated to analyse, and therefore an asymptotic reduction is required based on the fact that certain terms are small. The small terms are identified through nondimensionalisation. Howison [66] explains that any equation describing a physical process is a balance of the controlling mechanisms, but the terms are not of equal importance. The relative importance is assessed through nondimensionalisation. There are many advantages to working in nondimensional form, although there is a standard complaint that it may obscure the underlying physics. For this reason, it is often valuable (at least for the non-mathematical community) to revert to the dimensional form at the end of a calculation.

In the case of the Stefan problem, the nondimensional variables are related to their dimensional counterparts by the following relations:

$$\hat{x} = \frac{x}{L}, \qquad \hat{s} = \frac{s}{L}, \qquad \hat{t} = \frac{t}{\tau}, \qquad \hat{\theta} = \frac{\theta - T_m}{\Delta\theta}, \qquad \hat{T} = \frac{T - T_m}{\Delta T}, \tag{1.13}$$

where L, τ, $\Delta\theta$, ΔT are the appropriate length, time, and temperature scales. There is often some freedom of choice when nondimensionalising a problem. At times, the choice may be driven by prior knowledge of the physical situation or a particular requirement for the solution. The existence of large parameter values in an equation often indicates a poor scaling, which must then be re-worked. The existence of boundary layers, regions of rapid change in a variable, may require local rescaling. In a semi-infinite domain, there may be no natural choice for L and so it may be left arbitrary or motivated by a boundary condition. In a finite domain, L may be chosen as the material dimension or through some balance of terms. Time-scales may be chosen to highlight certain physical features: do we want to work on the time-scale of heat flow or phase change? With fluid flow, a further possible time-scale is introduced, the time taken for the fluid to traverse a certain distance. Perhaps we need to find the small time behaviour to provide an initial condition for a numerical solution? Nondimensionalisation is seldom straightforward but it is definitely rewarding when done correctly.

To illustrate the method, we now consider the case where the phase change is driven by heat flow through the solid, so we start by scaling the solid heat equation:

$$\frac{\rho_s c_s \Delta T}{\tau} \frac{\partial \hat{T}}{\partial \hat{t}} = \frac{k_s \Delta T}{L^2} \frac{\partial^2 \hat{T}}{\partial \hat{x}^2} . \tag{1.14}$$

The temperature scale cancels on either side, while balancing terms suggests that the time-scale $\tau = \rho_s c_s L^2 / k_s = L^2/\alpha_s$ depends on the length-scale L (where $\alpha_s = k_s/(\rho_s c_s)$ is the thermal diffusivity of the solid). This indicates that the time taken for heat flow varies with the square of the dimension. If phase change is driven by heat flow through the liquid, the choice $\tau = L^2/\alpha_l$ may be preferable; alternatively the time-scale may be chosen to balance terms in the Stefan condition. The current choice of time-scale leads to

$$\frac{\partial \hat{T}}{\partial \hat{t}} = \frac{\partial^2 \hat{T}}{\partial \hat{x}^2} . \tag{1.15}$$

Applying the same scale to the liquid heat equation results in

$$\frac{\partial \hat{\theta}}{\partial \hat{t}} = \alpha \frac{\partial^2 \hat{\theta}}{\partial \hat{x}^2} , \tag{1.16}$$

where the nondimensional group $\alpha = \alpha_l/\alpha_s = k_l \rho_s c_s/(k_s \rho_l c_l)$ is the ratio of liquid to solid thermal diffusivities. In many mathematical studies, it is assumed that physical properties are the same in either phase, and then we may simplify such that $\alpha = 1$. If flow is included, then Eq. (1.16) is replaced by

$$\frac{\partial \hat{\theta}}{\partial \hat{t}} + \frac{U\tau}{L} u \frac{\partial \hat{\theta}}{\partial \hat{x}} = \alpha \frac{\partial^2 \hat{\theta}}{\partial \hat{x}^2} . \tag{1.17}$$

When the flow is driven by the phase change, an obvious choice for the velocity scale is $U = L/\tau$, but this is not always the case (as will be spotted in Chap. 3 when the flow is driven by external forces). When $U \neq L/\tau$ but $\tau = L^2/\alpha_l$, then $U\tau/L = UL/\alpha_l$ may be recognised as the Péclet number, i.e. the ratio of advective to diffusive heat transport. A large Péclet number indicates that heat is primarily transferred by the flow; a small value indicates that diffusion is the dominant mechanism.

The Stefan condition (1.2) becomes

$$\rho_s L_m \frac{L}{\tau} \frac{d\hat{s}}{d\hat{t}} = \frac{k_s \Delta T}{L} \frac{\partial \hat{T}}{\partial \hat{x}} - \frac{k_l \Delta \theta}{L} \frac{\partial \hat{\theta}}{\partial \hat{x}} , \tag{1.18}$$

which may be rearranged to

$$\beta \frac{d\hat{s}}{d\hat{t}} = \frac{\partial \hat{T}}{\partial \hat{x}} - k \frac{\partial \hat{\theta}}{\partial \hat{x}} . \tag{1.19}$$

This expression contains two nondimensional groups, the ratio $k = k_l \Delta\theta/(k_s \Delta T)$ and the Stefan number. Now we pause momentarily to discuss the Stefan number since it will play an important role in the analysis throughout this book. Here it is defined as the ratio of latent heat to sensible heat. With the current choice of time-scale, based on heat flow through the solid, this leads to a Stefan number:

$$\beta = \frac{L_m}{c_s \Delta T} . \tag{1.20}$$

If the phase change were driven by heat flow through the liquid, the denominator should instead contain $c_l \Delta\theta$. The definition of the Stefan number (1.20) follows naturally from the nondimensionalisation of the Stefan condition. The same form is employed in numerous texts such as [40, 54, 89]. However, it is also often defined as the inverse of β, so providing a level of confusion in the community (see the Wikipedia entry for the Stefan number or Crank [36]). One researcher's large Stefan number solution may be another's small Stefan case. Beware! It may also be denoted St but we avoid this since it may be mistaken for the speed of the front s_t.

So far the temperature scale has not been clarified; again this is situation dependent. A typical choice for solidification would be $\Delta T = T_i - T_m > 0$, where T_i is the initial temperature of the fluid, for melting $\Delta T = T_m - T_i$. If the phase change is driven by heat exchange at the boundary, where the external temperature is T_a, then another possibility is $|T_a - T_m|$. The two temperature scales, ΔT, $\Delta\theta$, are often taken to be equal but, as we will see later, this is not necessarily a sensible choice.

One final point about nondimensionalisation. It is rather tedious to put hats on all nondimensional quantities, so the standard approach is to define the new variables and then immediately drop the hat notation. Consequently, the dimensional and nondimensional quantities appear the same. This issue is dealt with by applying common sense. If an equation has nondimensional numbers as coefficients or specified values (initial or at a boundary) are without units, then clearly, we have a nondimensional system. If the coefficients are dimensional or physical quantities and units are provided, then the system is dimensional. We will be gentle in this section, retaining the hats, but be warned they will be dropped faster than a hot potato or may never appear in subsequent chapters.

In Table 1.1, we present thermophysical properties for a variety of materials. These allow us to estimate the magnitude of the nondimensional numbers. In each case, only a single representative value is provided for each property. In reality, they may vary, for example, with pressure or temperature. Tin can exist in a number of different structural forms (allotropes), with the two main ones being α-tin and β-tin, which have very distinct properties (here we use the most common, β-tin). Clear

Table 1.1 Physical and thermodynamic parameter values for various materials. The values for σ_{sl} are taken from [19, 70, 128]

Substance	T_m (K)	L_m (J/Kg)	c_l, c_s (J/Kg·K)	ρ_l, ρ_s (kg/m^3)	k_l, k_s (W/m·K)	σ_{sl} (N/m)
Water	273.15	3.34×10^5	4181, 2050	1000, 920	0.55, 2.20	0.03
Gold	1337	6.37×10^4	163, 129	17,300, 19,300	106, 317	0.27
Lead	600	2.30×10^4	148, 128	10,700, 11,300	16,35	0.05
Tin	505	5.85×10^4	268, 230	6980, 7180	30, 67	0.064
Copper	136	2.09×10^5	530, 385	8020, 8960	157.2, 413	0.182
Salol	316	9×10^4	1580, -	1180, 1250	0.18, -	-
n-Octadecane	301.3	2.435×10^5	2161, 2151	771, 930	0.15, 0.372	-

or glaze ice has a density close to that of water, around 920 kg/m^3, while rime ice that contains pockets of air may be around 880 kg/m^3. Be careful when you assign parameter values. The parameter σ_{sl} represents the surface tension between the solid and liquid phases of the material.

1.3.1 Choice of Time-Scale

In practical applications, the Stefan number is often large. If we consider the physical values shown in Table 1.1, we find $L_m/c_s \approx 162, 493, 179, 254$ K for water, gold, lead, and tin, respectively. This means that if we choose a temperature scale ΔT with exactly these values, then $\beta = L_m/(\Delta T c_s) = 1$ and the Stefan condition is well-scaled. Of course, we are unlikely to use such high values to effect a phase change. For water, a more typical value would be $\Delta T = 5$ K, in which case $\beta \approx 30$. In fact, in most practical examples, we will find $\beta \gg 1$.

The observation concerning the size of β has consequences for the analysis of the model. If we now consider Eq. (1.19), then

$$\frac{d\hat{s}}{d\hat{t}} = \frac{1}{\beta}\left(\frac{\partial \hat{T}}{\partial \hat{x}} - k\frac{\partial \hat{\theta}}{\partial \hat{x}}\right). \tag{1.21}$$

As $\beta \to \infty$, this indicates that $\hat{s}_{\hat{t}} \to 0$, and the velocity of the phase change front appears negligible, suggesting that the phase change process is also negligible. We have lost the very effect that we were planning to study! This could make for a very disappointing start to a book. The physical meaning behind this observation is simple. We have chosen a time-scale $\tau = L^2/\alpha_s$ based on the rate of heat flow through the solid, but there is another important time-scale, that of the phase change. The Stefan number first appeared in the nondimensional Stefan condition (1.19), and when we substituted for the solid diffusion time-scale, this led to $\beta = L_m/(c_s \Delta T)$. If we nondimensionalise the Stefan condition, (1.2), and now leave the time-scale unspecified, then the coefficient on the left-hand side is

$$\beta = \frac{\rho_s L_m L^2}{\tau k_s \Delta T} . \tag{1.22}$$

We now see that a large Stefan number may be interpreted either in terms of a short time-scale, $\tau \ll 1$, or small temperature scale, $\Delta T \ll 1$. Using the short time-scale interpretation indicates that heat flow is much more rapid than the phase change process. If we actually want to focus on the phase change, it is then more sensible to choose the phase change time-scale $\tau = \rho_s L_m L^2/(k_s \Delta T)$, rather than the more familiar thermal diffusion scale. This converts the nondimensional Stefan condition to

$$\frac{d\hat{s}}{d\hat{t}} = \frac{\partial \hat{T}}{\partial \hat{x}} - k\frac{\partial \hat{\theta}}{\partial \hat{x}} , \tag{1.23}$$

and now all terms balance. We have successfully reintroduced the effect to be studied. The heat equations become

$$\frac{1}{\beta}\frac{\partial \hat{T}}{\partial \hat{t}} = \frac{\partial^2 \hat{T}}{\partial \hat{x}^2} , \qquad \frac{1}{\beta}\frac{\partial \hat{\theta}}{\partial \hat{t}} = \alpha\frac{\partial^2 \hat{\theta}}{\partial \hat{x}^2} . \tag{1.24}$$

In the limit $\beta \to \infty$ this indicates

$$\frac{\partial^2 \hat{T}}{\partial \hat{x}^2} , \frac{\partial^2 \hat{\theta}}{\partial \hat{x}^2} \to 0 , \tag{1.25}$$

which lead to linear profiles in the solid and liquid phases. The fact that the time derivatives in the heat equation are negligible indicates that if we work on the time-scale of the phase change (which is much slower than the heat flow time-scale), then the temperature has time to equilibrate to its steady-state value. The formulation with the left-hand sides of Eq. (1.24) set to zero is termed the pseudo- or quasi-steady-state problem. It does not constitute a true steady-state since the domain over which the heat equations apply changes with time. In the following chapter, we will investigate this scaling further when considering perturbation solutions. The linear form for the temperature will appear frequently, and the pseudo-steady state may be identified as the leading order term in a series expansion in powers of $1/\beta$ for $\beta \gg 1$.

1.4 Conclusion

The basic thermal Stefan problem has now been defined. It consists of heat equations in both phases linked by an energy balance to track the interface as well as suitable boundary conditions. Nondimensionalisation indicates which terms are dominant and which are negligible (we will exploit this technique in all subsequent chapters).

The assumptions imposed in producing the basic Stefan problems introduced in the current chapter restrict their applicability. The goal of this book is to investigate physically realistic Stefan problems, and this requires the removal of some restrictions at least and will lead to more complex systems than the ones presented so far. The removal of restrictions will be the topic of later chapters, but first, to be able to deal with the complex systems, we need to understand basic solution forms and methodologies. So, in the following chapter, we present exact and approximate solution methods for the standard one-dimensional Stefan problem.

Chapter 1 Exercises

1. (a) Nondimensionalise the one-dimensional heat equation:

$$\rho c \frac{\partial T}{\partial t} = k \frac{\partial^2 T}{\partial x^2}$$

to determine the time-scale for a given length-scale.

(b) Now estimate the time-scale for heat flow through solid and liquid gold and solid and liquid water (assuming there is no flow) for a sample of 10 cm (parameter values are provided in Table 1.1).

Note, in both cases, the liquid time-scale is significantly longer – demonstrating that solids conduct heat much more rapidly than liquids (however, liquids may also move heat through convection or advection, which we neglect here). Also, the metal is a much better conductor than water. Feel free to try other materials.

2. If a phase change is driven by heat flow through a liquid, such that the chosen time-scale is $\tau = \rho_l c_l L^2/k_l$. Write down the expression for the Stefan number.
3. Consider the Newton cooling/Robin condition of Eq. (1.5):

$$k \frac{\partial T}{\partial x} = \mathscr{H}(T_a - T) .$$

This is the most general of the boundary conditions discussed in this chapter since it may reduce to the other forms in the correct limits.
Defining $\hat{T} = (T - T_m)/(T_a - T_m)$ and $\hat{x} = x/L$, write the boundary condition in nondimensional form.
The Nusselt number $Nu = \mathscr{H}L/k$ defines the ratio of surface heat transfer to that conducted from the surface. For what Nu limits do we obtain the Dirichlet condition and the insulated condition $\hat{T}_{\hat{x}} = 0$?
Choose appropriate re-scalings to obtain the Neumann condition $\hat{T}_{\hat{x}} = \text{constant}$.
4. For melting driven by a temperature gradient through the solid phase, calculate the Stefan number $\beta = L_m/(c_s \Delta T)$ for n-octadecane, gold, and ice. From this, can you estimate which will melt the slowest/fastest?

If not, can you suggest a more reliable measure of the melting rate?

5. There are many methods for storing energy; one way is to use phase change materials. These absorb energy when melting and so can be used as a coolant in warm environments. Materials such as n-octadecane are frequently employed. Why?
6. Consider a slab of tin initially at 450 K. It is subsequently heated at one end to a temperature of 600 K. If we scale such that the solid temperature $\hat{T} \in [-1, 0]$ and the liquid temperature $\hat{\theta} \in [0, 1]$, what nondimensionalisation has been employed?
7. The first example studied in the following chapter is the nondimensional system:

$$\frac{\partial T}{\partial t} = \frac{\partial^2 T}{\partial x^2} \qquad 0 < x < s$$

$$\frac{\partial \theta}{\partial t} = \alpha \frac{\partial^2 \theta}{\partial x^2} \qquad x > s$$

$$\beta \frac{ds}{dt} = \frac{\partial T}{\partial x} - k \frac{\partial \theta}{\partial x} \qquad x = s\,,$$

where the hat notation has been dropped. These are subject to

$$T(0, t) = T_0\,, \quad T(s, t) = \theta(s, t) = 0\,, \quad \theta|_{x\to\infty} = \theta(x, 0) = 1\,,$$

where $T_0 < 0$, $\Delta T = \Delta\theta = T_i - T_m$, $\theta(x, 0) = T_i$ is the constant initial liquid temperature, and $\beta = L_m/(c_s(T_i - T_m))$. Write down the relation between the dimensional and nondimensional temperatures and then the corresponding dimensional system.

8. In Sect. 2.1.4, the one-phase growth of a spherical nanocrystal from a colloidal solution is considered. This is governed by

$$\frac{\partial c}{\partial t} = \frac{D}{r^2}\frac{\partial}{\partial r}\left(r^2 \frac{\partial c}{\partial r}\right), \qquad r > r_p(t)\,,$$

$$\frac{dr_p}{dt} = V_m D \left.\frac{\partial c}{\partial r}\right|_{r_p},$$

where c is the concentration of solute surrounding the crystal, $r_p(t)$ the sphere radius, D the diffusion coefficient, and V_m the ratio of molar mass of the concentrate to the crystal density. If the initial concentration $c(r, 0) = c_i$ is constant and at the interface $c(r_p, t) = c_p$, nondimensionalise the system such that the time-scale is taken from the diffusion equation and

$$\hat{c}|_{\hat{r}\to\infty} = 0\,, \quad \hat{c}(\hat{r}_p, \hat{t}) = -1\,, \quad \hat{c}(\hat{r}, 0) = 0\,.$$

What is the equivalent to the Stefan number? (Note, there is no natural length-scale so choose an undefined constant.)

9. In Sect. 3.2, solidification in a microchannel is analysed. A fluid flows between two cold walls located at $z = \pm H$ such that a solid builds up on the wall. The new solid is located between $z \in \pm s$. This acts to restrict the flow. The basic system is described by the two-dimensional, Cartesian form of Eqs. (1.6a, 1.6c 1.7), where the unit normal $\hat{\mathbf{n}} = (-s_x, 1)/\sqrt{1+s_x^2}$. These are subject to

$$\theta(x,z,0) = T_0\,, \qquad \theta(x,s,t) = T_m\,, \qquad \left.\frac{\partial \theta}{\partial z}\right|_{z=0} = 0\,,$$

$$T(x,s,t) = T_m\,, \qquad T(x,H,t) = T_w\,,$$

where the wall temperature $T_w < T_m < T_0$. The system is written in nondimensional form using

$$\hat{T} = \frac{T - T_m}{T_m - T_w}\,, \qquad \hat{\theta} = \frac{\theta - T_m}{T_0 - T_m}\,, \qquad \hat{x} = \frac{x}{L}\,, \qquad \hat{z} = \frac{z}{H}\,,$$

$$\hat{s} = \frac{s}{H}\,, \qquad \hat{t} = \frac{t}{\tau}\,.$$

Write the nondimensional form of the heat equations and Stefan condition (this requires nondimensionalising the unit normal as well).

Noting that $\epsilon = H/L \ll 1$, infer the reduced form of governing equations where terms of $\mathcal{O}(\epsilon^2)$ are neglected.

Chapter 2
Exact and Approximate Solutions

There exists a plethora of exact solutions to Stefan problems. Unfortunately, very few of them are of any practical use, involving unphysical situations and linked to boundary conditions chosen to match the solution form (rather than being practically motivated). Hill [64] laments that, after a hundred years of research, solutions of the type presented by Neumann are the only physically realistic examples, while other exact solutions are not of genuine practical interest. The true state is not quite so lamentable.

In this chapter, we begin by providing a description of the few practically useful solutions (to be fair, even these have strong limitations). As stated in the previous chapter, the first of these, the Neumann solution, was presented in a series of lectures in the 1860s and so predates the Stefan problem itself [36]. It requires a fixed temperature (Dirichlet) boundary condition, is derived using the similarity variable $\eta = x/\sqrt{t}$, and provides a solution in the form of error functions. As stated in Carslaw and Jaeger [29], the form $\text{erf}(x/\sqrt{t})$ is the only solution based purely on the grouping $x/\sqrt{t}$. It may be used to determine the Neumann solution, which satisfies Dirichlet boundary conditions, but is incapable of satisfying Neumann (ironically) or Robin conditions.

We begin by detailing the Neumann solution, and related forms where an exact solution is possible, before moving on to approximate solution techniques. Many of these will be employed in subsequent chapters, so don't skip this chapter!

T. G. Myers, *Theoretical and Practical Stefan Problems*, Interdisciplinary Applied Mathematics 63, https://doi.org/10.1007/978-3-032-04826-4_2

2.1 Exact Solutions

2.1.1 The Neumann Solution for One-Dimensional Solidification

As a first example, we choose to analyse one of the most basic forms of Stefan problem, namely, solidification of a semi-infinite liquid driven by imposing a fixed temperature (below the phase change temperature) at the left-hand boundary. The liquid and solid densities are equal and constant. The configuration is shown in Fig. 1.1. The nondimensional mathematical model is defined by Eqs. (1.15, 1.16, 1.19)

$$\frac{\partial T}{\partial t} = \frac{\partial^2 T}{\partial x^2} \qquad 0 < x < s \tag{2.1a}$$

$$\frac{\partial \theta}{\partial t} = \alpha \frac{\partial^2 \theta}{\partial x^2} \qquad x > s \tag{2.1b}$$

$$\beta \frac{ds}{dt} = \frac{\partial T}{\partial x} - k \frac{\partial \theta}{\partial x} \qquad x = s\,, \tag{2.1c}$$

(where we have dropped the hat notation) and are to be solved subject to

$$T(0,t) = T_0\,, \quad T(s,t) = \theta(s,t) = 0\,, \quad \theta|_{x\to\infty} = \theta(x,0) = 1\,. \tag{2.1d}$$

To obtain this system, we have employed the nondimensionalisation of Eq. (1.13) with temperature scales $\Delta T = \Delta\theta = T_i - T_m$ where $T_i = \theta|_{t=0}$ is the constant initial liquid temperature (see Chap. 1, Exercise 7). With the same temperature scaling in both phases, we obtain a non-zero boundary temperature, $T(0,t) = T_0 < 0$. The Stefan number is $\beta = L_m/(c_s(T_i - T_m))$, $\alpha = \alpha_l/\alpha_s$, and $k = k_l/k_s$.

Neumann's solution is obtained by working with the similarity variable $\eta = x/\sqrt{t}$. Defining new temperature functions $T(x,t) = F_1(\eta)$ and $\theta(x,t) = F_2(\eta)$, the derivatives become

$$\frac{\partial T}{\partial t} = \frac{dF_1}{d\eta}\frac{\partial \eta}{\partial t} = \frac{dF_1}{d\eta}\left(-\frac{x}{2t^{3/2}}\right) = \frac{dF_1}{d\eta}\left(-\frac{\eta}{2t}\right), \tag{2.2a}$$

$$\frac{\partial T}{\partial x} = \frac{dF_1}{d\eta}\frac{\partial \eta}{\partial x} = \frac{1}{\sqrt{t}}\frac{dF_1}{d\eta}, \tag{2.2b}$$

$$\frac{\partial^2 T}{\partial x^2} = \frac{\partial}{\partial x}\left(\frac{1}{\sqrt{t}}\frac{dF_1}{d\eta}\right) = \frac{1}{\sqrt{t}}\frac{\partial \eta}{\partial x}\frac{\partial}{\partial \eta}\left(\frac{dF_1}{d\eta}\right) = \frac{1}{t}\frac{d^2 F_1}{d\eta^2}, \tag{2.2c}$$

and similar for θ.

The solid temperature now satisfies

$$-\left(\frac{\eta}{2}\right)\frac{dF_1}{d\eta}=\frac{d^2F_1}{d\eta^2}\,, \tag{2.3}$$

which is independent of x and t (if either remains, then the reduction has failed). Writing $\partial F_1/\partial\eta = G_1$ leads to

$$\frac{1}{G_1}\frac{dG_1}{d\eta}=-\frac{\eta}{2} \quad\Rightarrow\quad G_1=\frac{dF_1}{d\eta}=A'e^{-\eta^2/4}\,, \tag{2.4}$$

for some constant of integration A'. Integrating and applying $F_1(0) = T_0$ (since $x = 0$ corresponds to $\eta = 0$) determines

$$F_1 = T_0 + A\,\mathrm{erf}\left(\frac{\eta}{2}\right) \quad\Rightarrow\quad T = T_0 + A\mathrm{erf}\left(\frac{x}{2\sqrt{t}}\right), \tag{2.5}$$

where the constant $A = \sqrt{\pi}A'$. A similar analysis in the liquid, subject to $F_2(\eta) \to 1$ as $\eta \to \infty$ (corresponding to $\theta|_{x\to\infty} \to 1$), leads to

$$F_2 = 1 - B\mathrm{erfc}\left(\frac{\eta}{2\sqrt{\alpha}}\right) \quad\Rightarrow\quad \theta = 1 - B\mathrm{erfc}\left(\frac{x}{2\sqrt{\alpha t}}\right), \tag{2.6}$$

with B constant. At the moving interface, $x = s(t)$, the temperatures take on the phase change value $T = \theta = 0$; hence,

$$0 = T_0 + A\,\mathrm{erf}\left(\frac{s}{2\sqrt{t}}\right) = 1 - B\,\mathrm{erfc}\left(\frac{s}{2\sqrt{\alpha t}}\right). \tag{2.7}$$

Since A and B are constant, these equations can only be valid if there is no t dependence, which requires $s \propto \sqrt{t}$. This is the infamous square root dependence first predicted by Lamé and Clapeyron. Setting $s = 2\lambda\sqrt{t}$ (with λ constant) determines

$$A = -\frac{T_0}{\mathrm{erf}\lambda} \qquad B = \frac{1}{\mathrm{erfc}(\lambda/\sqrt{\alpha})}\,. \tag{2.8}$$

The temperatures are then

$$F_1 = T_0\left(1 - \frac{\mathrm{erf}\ (\eta/2)}{\mathrm{erf}\ (\lambda)}\right) \quad\Rightarrow\quad T = T_0\left(1 - \frac{\mathrm{erf}\ \left(x/2\sqrt{t}\right)}{\mathrm{erf}\ (\lambda)}\right), \tag{2.9}$$

$$F_2 = 1 - \frac{\mathrm{erfc}\ \left(\eta/(2\sqrt{\alpha})\right)}{\mathrm{erfc}(\lambda/\sqrt{\alpha})} \quad\Rightarrow\quad \theta = 1 - \frac{\mathrm{erfc}\ \left(x/(2\sqrt{\alpha t})\right)}{\mathrm{erfc}(\lambda/\sqrt{\alpha})}\,, \tag{2.10}$$

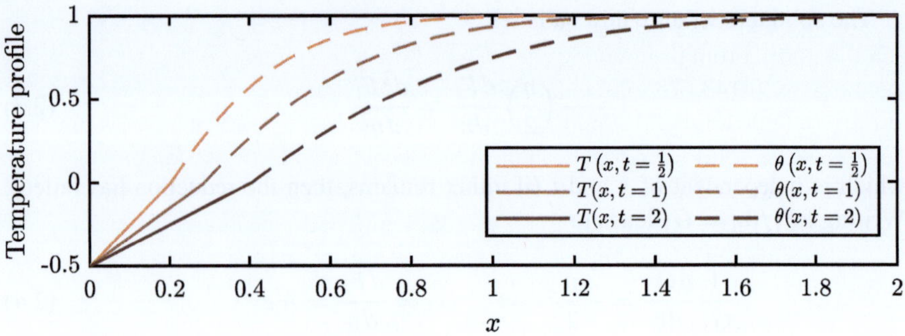

Fig. 2.1 Temperature profiles for two-phase solidification at times $t = 0.5, 1, 2$, with $T_0 = -0.5$, $k = 0.227$, $\alpha = 0.113$, $\beta = 8$ (which, from Eq. (2.12), determines $\lambda = 0.143$) solid lines represent $T(x, t)$, and dashed lines represent $\theta(x, t)$. As time increases, the interface moves between $s = 0.202$, 0.286, and 0.405

Noting that

$$\frac{\partial}{\partial x}\left(\operatorname{erf}\left(\frac{x}{2\sqrt{t}}\right)\right) = -\frac{\partial}{\partial x}\left(\operatorname{erfc}\left(\frac{x}{2\sqrt{t}}\right)\right) = \frac{\exp(-x^2/(4t))}{\sqrt{\pi t}}. \tag{2.11}$$

the Stefan condition (2.1c), which is applied at $x = s = 2\lambda\sqrt{t}$, may be rearranged to give the following transcendental equation for λ:

$$\beta\lambda = -\frac{1}{\sqrt{\pi}}\left[\frac{T_0 e^{-\lambda^2}}{\operatorname{erf}(\lambda)} + \frac{k e^{-\lambda^2/\alpha}}{\sqrt{\alpha}\operatorname{erfc}(\lambda/\sqrt{\alpha})}\right]. \tag{2.12}$$

The fact that the t dependence has dropped out is crucial to the statement that λ is constant. This indicates that a solution of the form $s = 2\lambda\sqrt{t}$ is indeed possible.

If we specify the material parameters α, β, k and the temperature driving the solidification, T_0, then any root finding code can easily solve equation (2.12) for the unknown λ. Once λ is determined, the interface position, $s = 2\lambda\sqrt{t}$, constants A, B, and hence temperature profiles are all known. The remarkable feature of this solution is that a problem involving two partial differential equations specified over an unknown, time-dependent domain has been reduced to a single transcendental equation for just one variable. This is an impressive simplification.

In Fig. 2.1, a typical solution is presented. Parameter values appropriate to the freezing of water are employed, with initial temperature 20 °C (293.15 K) and wall temperature −10 °C (263.15 K). Using the values from Table 1.1, this determines

$$\hat{T}_0 = \frac{T_0 - T_m}{T_i - T_m} = \frac{-10}{20} = -0.5\,, \qquad k = \frac{k_l}{k_s} = \frac{0.55}{2.2} = 0.25\,, \tag{2.13}$$

$$\alpha = \frac{k_l}{\rho_l c_l}\frac{\rho_s c_s}{k_s} = 0.113\,, \qquad \beta = \frac{L_m}{c_s(T_i - T_m)} \approx 8\,, \tag{2.14}$$

(note we have brought back the hat on $\hat{T}_0$ to distinguish it from the dimensional temperature). From the numerical solution of the transcendental equation (2.12), we obtain $\lambda = 0.143$. The three sets of lines in the figure represent the temperature profile in both solid and liquid components at times $t = 0.5, 1, 2$. The solid lines represent the solid temperature, and the dashed lines the liquid temperature. The interface is located at $s = 2 \times 0.143 \times \sqrt{t} = 0.202, 0.286, 0.405$ for $t = 0.5, 1, 2$, respectively. It is worth noting that the solid temperature is approximately linear at all times shown in the figure: this linear form will recur in a number of solutions presented in subsequent sections.

In the previous chapter, when discussing the kinetic energy, we mentioned the issue of an infinite initial velocity. This may now be quantified, since $s = 2\lambda\sqrt{t}$; then the front velocity $s_t = \lambda/\sqrt{t}$, which obviously tends to infinity as $t \to 0$. The infinite velocity is an obvious consequence of the mismatch between the initial and boundary temperatures. The initial condition states that the liquid temperature is everywhere above zero, the boundary condition is fixed at $T_0 < 0$, and at the interface, $T = \theta = 0$. At small times, conduction through the solid dominates, so the Stefan condition indicates the velocity $s_t \propto -T_x$, and for sufficiently small δx, we may approximate the temperature gradient $T_x \approx \delta T/\delta x \approx (T_0 - 0)/\delta x$. As $t \to 0$, the distance over which the temperature jump takes place $\delta x \to 0$ and so $T_x \to -\infty$. Physically, this is an indication that the standard conditions are unrealistic: it is impossible to have a uniform temperature throughout a material and then instantaneously change to a distinct value at a single point. For this reason, the fixed temperature boundary condition should, in general, be thought of as an approximation (or simply unrealistic?) employed purely for mathematical convenience. The approximation may be reasonable at large times, but as evidenced by the initial infinite velocity, is rather poor as $t \to 0$. Students may recall difficulties in solving the heat equation using separation of variables when there is a jump between the boundary and initial temperatures. The oscillations that always occur at small times (irrespective of the number of terms in the series) are an example of *Gibbs' phenomenon*, which results from trying to describe a jump with a series of continuous functions. Unphysical conditions often lead to mathematical problems.

2.1.1.1 Extension of Neumann's Solution to Different Densities

With a density change at the interface, a liquid velocity is induced:

$$u = \left(1 - \frac{\rho_s}{\rho_l}\right)\frac{ds}{dt}. \tag{2.15}$$

The calculation of the above interface velocity is detailed in Chap. 3 (see Eq. (3.19)), for now just trust me. Equation (2.15) simply states that, assuming the solid is fixed, the liquid must move due to the density change upon solidification. If the density of the solid is lower than that of the liquid, then the solid must expand into the liquid, which is then forced to the right and u is positive. If the solid has a higher density, then contraction occurs during the phase change and liquid moves to fill up the vacated space, and u is then negative.

The liquid motion does not affect the heat flow in the solid, which remains identical to that of the previous section. In the liquid, the heat equation (1.17) (with velocity scale $U = L/\tau$) must now account for advection, leading to

$$\frac{\partial \theta}{\partial t} + \left(1 - \frac{\rho_s}{\rho_l}\right) \frac{ds}{dt} \frac{\partial \theta}{\partial x} = \alpha \frac{\partial^2 \theta}{\partial x^2} . \tag{2.16}$$

With the similarity variable $\eta = x/\sqrt{t}$ and assuming $s = 2\lambda\sqrt{t}$, for some constant λ, we obtain

$$-\left[\frac{\eta}{2} - \left(1 - \frac{\rho_s}{\rho_l}\right)\lambda\right] \frac{dF_2}{d\eta} = \alpha \frac{d^2 F_2}{d\eta^2} , \tag{2.17}$$

which is independent of x, t. As before, we introduce $G_2 = dF_2/d\eta$ and integrate

$$G_2 = \frac{dF_2}{d\eta} = B' \exp\left(-\frac{\eta^2}{4\alpha} + \left(1 - \frac{\rho_s}{\rho_l}\right)\frac{\lambda\eta}{\alpha}\right) , \tag{2.18}$$

where B' is a constant of integration. Integrating a second time

$$F_2 = 1 - B \exp\left[\left(1 - \frac{\rho_s}{\rho_l}\right)^2 \frac{\lambda^2}{\alpha}\right] \operatorname{erfc}\left[\frac{\eta}{2\sqrt{\alpha}} - \left(1 - \frac{\rho_s}{\rho_l}\right)\frac{\lambda}{\sqrt{\alpha}}\right] . \tag{2.19}$$

The second constant of integration has been calculated by applying the boundary condition at infinity, $F_2 \to 1$ and for simplicity, we write $B = \sqrt{\pi\alpha}B'$. Applying the temperature condition at the interface, $F_2(2\lambda) = 0$, indicates

$$B = \left(\exp\left[\left(1 - \frac{\rho_s}{\rho_l}\right)^2 \frac{\lambda^2}{\alpha}\right] \operatorname{erfc}\left[\frac{\lambda}{\sqrt{\alpha}}\frac{\rho_s}{\rho_l}\right]\right)^{-1} , \tag{2.20}$$

and so

$$F_2 = 1 - \operatorname{erfc}\left[\frac{\eta}{2\sqrt{\alpha}} - \left(1 - \frac{\rho_s}{\rho_l}\right)\frac{\lambda}{\sqrt{\alpha}}\right] \Big/ \operatorname{erfc}\left[\frac{\lambda}{\sqrt{\alpha}}\frac{\rho_s}{\rho_l}\right] . \tag{2.21}$$

This equation holds for both cases $\rho_s < \rho_l$ and $\rho_s > \rho_l$. By calculating the temperature gradients and setting $\eta = 2\lambda$, the Stefan condition, Eq. (2.1c), may be written as

$$\beta\lambda = -\frac{1}{\sqrt{\pi}}\left[\frac{T_0 e^{-\lambda^2}}{\mathrm{erf}\,(\lambda)} + \frac{k e^{-\lambda^2\rho_s^2/(\alpha\rho_l^2)}}{\sqrt{\alpha}\,\mathrm{erfc}\left(\lambda\rho_s/(\sqrt{\alpha}\rho_l)\right)}\right]. \tag{2.22}$$

Comparison with Eq. (2.12) shows that the effect of the density change is manifested through the second term on the right-hand side. This is a result of the temperature gradient in the liquid. In the limit $\rho_s/\rho_l = 1$, Eq. (2.22) reduces to the zero velocity case, Eq. (2.12).

The above solution is a generalisation of that presented in [29] for the case where $\rho_s > \rho_l$. Crank [35, §3.2] quotes the solution of [29] and refers to the different density case as the "commonly neglected problem". In Chap. 4, we will see that although this solution improves on the classical Neumann solution, it still relies on a significant approximation: the induced fluid flow introduces kinetic energy into the Stefan condition, which is not accounted for in the above formulation.

2.1.2 Evolution of a Semi-infinite Material with a Mushy Zone

Solomon et al. [144] propose an idealised model for a semi-infinite mushy region. Initially, the liquid is at the phase change temperature, when the base $x = 0$ is immediately changed to a temperature below this value. Subsequently, solid occupies the region $x \in [0, s_1(t)]$; the region $[s_1(t), s_2(t)]$ is occupied by a mixed layer of solid and liquid, while $x \geq s_2(t)$ is pure liquid. The mixed, or mushy, region is assumed to have a constant volume fraction θ of solid particles (hence, $(1-\theta)$ liquid). A second (very) significant assumption is that the width of the mushy layer is assumed inversely proportional to the temperature gradient:

$$T_x(s_1, t)(s_2 - s_1) = \gamma \tag{2.23}$$

where the constant γ is determined experimentally. This restrictive condition is inspired by experimental observations where, for paraffin wax, $\gamma = 0.031$ K.

In nondimensional form, we may write down the system:

$$\frac{\partial T}{\partial t} = \frac{\partial^2 T}{\partial x^2} \qquad 0 < x < s_1 \tag{2.24a}$$

$$\beta\left[\theta\frac{ds_1}{dt} + (1-\theta)\frac{ds_2}{dt}\right] = \left.\frac{\partial T}{\partial x}\right|_{x=s_1}, \tag{2.24b}$$

subject to

$$T(0,t) = -1\ , \quad T(s_1,t) = 0\ . \tag{2.24c}$$

The mush and water layers are at the phase change temperature for all time, so no heat equations are required.

Following the analysis of Sect. 2.1.1, we define the similarity variable, assume $s_1 = 2\lambda\sqrt{t}$, and determine

$$T = -1 + \frac{\operatorname{erf}(x/(2\sqrt{t}))}{\operatorname{erf}\lambda} \qquad \left.\frac{\partial T}{\partial x}\right|_{x=s_1} = \frac{\exp(-\lambda^2)}{\sqrt{\pi t}\,\operatorname{erf}\lambda}\,. \tag{2.25}$$

From Eq. (2.23), it follows that s_2 must also take the square root form $s_2 = 2\mu\sqrt{t}$ and consequently

$$\frac{2\exp(-\lambda^2)}{\sqrt{\pi}\,\operatorname{erf}\lambda}(\mu - \lambda) = \gamma \quad \Rightarrow \quad \mu = \lambda + \frac{\gamma\sqrt{\pi}\,\operatorname{erf}\lambda}{2\exp(-\lambda^2)}\,. \tag{2.26}$$

This defines the relation between μ and λ. Finally, the Stefan condition gives

$$\beta\left[\theta\lambda + (1-\theta)\mu)\right] = \frac{\exp(-\lambda^2)}{\sqrt{\pi}\,\operatorname{erf}\lambda}\,. \tag{2.27}$$

Equation (2.26) may be used to eliminate μ, and we are left with a transcendental equation to determine λ for given values of θ and γ. Once λ is found, all other quantities follow.

This provides one of the few exact solutions for phase change with a mushy zone. It requires a highly restrictive assumption that the width of the mushy region is constant. However, in a field where solutions are scarce, any analytical progress is welcome.

2.1.3 *One-Phase Stefan Problems*

The one-phase Stefan problem is a standard approximation primarily studied to make the analysis simpler and give a basic understanding of the phase change process. In this book, we will often use one-phase approximations since it significantly reduces the number and length of equations. In general, the extension to two-phase is straightforward. Physically, a one-phase problem can correspond to situations where one of the material components is everywhere close to the phase change temperature or where one component simply disappears from the calculation. This is not as unlikely as it sounds, for example, in studies of sublimation (the direct transition from solid to gas or the related ablation), where the heat source is so great that a solid turns immediately to vapour, one may assume that the vapour (which tends to be blasted away) plays only a negligible role in the subsequent process. This is dealt with in studies of laser ablation or the slow destruction of heat shields on space vehicles as they enter the atmosphere [6, 87]. With crystal growth from a liquid solution, the position of the interface is controlled solely by the local concentration of the liquid, and so the crystal only enters into the problem as it defines the position of the boundary. This will be investigated in Sects. 2.1.4 and 5.3. One-phase reductions may be more formally viewed as leading order reductions in the limit of either large or small $k = k_l \Delta\theta/(k_s \Delta T)$; see [46, 106]. This is discussed further in Sect. 4.5.

To allow comparison with the results of the previous section, we will now consider the solidification of a semi-infinite liquid, which is initially at the phase change temperature (in the liquid state, this is termed the *liquidus*; a solid initially at the phase change temperature is at the *solidus*). Cooling is achieved by fixing the boundary $x = 0$ to a constant temperature below the phase change value. We begin by slightly adjusting the nondimensionalisation. In the two-phase problem, there were three distinct temperatures: the initial value, the boundary value, and the phase change value. Consequently, we were unable to scale out all temperatures. In the current situation, the initial and phase change values are the same, so a sensible nondimensionalisation takes the form

$$\hat{T} = \frac{T - T_m}{T_m - T_0} . \tag{2.28}$$

In this way, the nondimensional temperature ranges from -1 to 0. The mathematical problem may now be stated as

$$\frac{\partial T}{\partial t} = \frac{\partial^2 T}{\partial x^2} , \qquad 0 < x < s \tag{2.29a}$$

$$\beta \frac{ds}{dt} = \left. \frac{\partial T}{\partial x} \right|_{x=s} , \tag{2.29b}$$

subject to

$$T(0,t) = -1 \qquad T(s,t) = 0 \qquad s(0) = 0 . \tag{2.29c}$$

Since the solid does not exist at $t = 0$, there is no initial condition on the temperature.

The Neumann solution readily adapts to the one-phase problem. By simply neglecting the liquid contribution to Eq. (2.12) (or equivalently by setting the liquid conductivity to zero, $k_l = k = 0$) and setting $T_0 = -1$ in Eqs. (2.9, 2.12), we find

$$T = -1 + \frac{\operatorname{erf}\left(\frac{x}{2\sqrt{t}}\right)}{\operatorname{erf}(\lambda)} , \qquad \beta\sqrt{\pi}\lambda e^{\lambda^2} \operatorname{erf}(\lambda) = 1 , \tag{2.30}$$

where $s = 2\lambda\sqrt{t}$. Once the transcendental equation is solved, the temperature and interface position are known. Results are shown for the freezing of water with $\beta = 8$, in Fig. 2.2. Although the temperature scale is different to that of Fig. 2.1, the

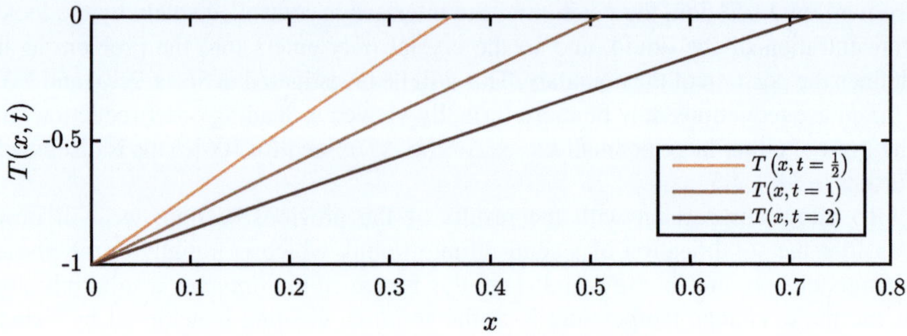

Fig. 2.2 Temperature profiles $T(x,t)$ for the one-phase solidification problem at times $t = 0.5, 1, 2$, with $T_0 = -1$, $\beta = 8$ (solving equation (2.30), this leads to $\lambda = 0.255$). As time increases, the interface moves between $s = 0.36, 0.51$, and 0.72

length-scale is the same (hence the position of the interfaces in the two figures are comparable), and it is clear that the interface in the one-phase problem advances much faster. The explanation for this is that in the two-phase case, the warmth of the liquid acts to slow down the freezing process. Here the liquid has no influence and so faster freezing ensues. The change in solidification rate may be quantified through the values of λ: in the two-phase problem, we obtained $\lambda = 0.143$, and for the equivalent one-phase problem, we obtain $\lambda = 0.245$, indicating an increase in freezing rate by a factor of close to 2. Again we observe that the temperature is approximately linear in the solid.

2.1.4 Spherically Symmetric One-Phase Growth

A problem that will be discussed in Chap. 5 concerns the growth of a nanocrystal from a colloidal solution. This is a chemical diffusion process that is mathematically equivalent to a thermal Stefan problem. Here we study a closely related problem, the growth of a sphere due to a flux of concentrate that occupies the surrounding liquid. The mathematical description of the one-phase problem is

$$\frac{\partial c}{\partial t} = \frac{1}{r^2}\frac{\partial}{\partial r}\left(r^2\frac{\partial c}{\partial r}\right) = \frac{\partial^2 c}{\partial r^2} + \frac{2}{r}\frac{\partial c}{\partial r}, \qquad r > r_p(t), \tag{2.31a}$$

$$\beta\frac{dr_p}{dt} = \left.\frac{\partial c}{\partial r}\right|_{r_p}, \tag{2.31b}$$

where c is the concentration of solute and $r_p(t)$ is the sphere radius. The first of these equations represents diffusion in the liquid surrounding the sphere. Equation (2.31b) is equivalent to a one-phase Stefan condition, but here derives from a mass balance at the boundary of the sphere; the volume increases due to the flux of solute to the surface. The 'Stefan number' is related to the molecular volume and mass diffusion. The above equations are to be solved subject to

$$c|_{r\to\infty} = 0\,, \quad c(r_p, t) = -1\,, \quad c(r, 0) = 0\,, \quad r_p(0) = 0\,. \tag{2.32}$$

The negative nondimensional concentration is simply due to the scaling. The far-field concentration must be higher than that at the crystal boundary since this is what drives the diffusion towards the crystal.

In the actual nanocrystal growth problem, the concentration at the boundary r_p varies and growth starts from a seed crystal, $r_p(0) \neq 0$. However, that system does not permit an analytical solution, so here we investigate equations (2.31, 2.32), which do permit an exact solution and then provide a guide for the nanocrystal growth, which must be solved approximately.

Again we use a similarity reduction with the variable $\eta = r/\sqrt{t}$ and define a new concentration function:

$$F(\eta) = c(r, t) \tag{2.33}$$

such that

$$\frac{\partial c}{\partial t} = \frac{dF}{d\eta}\left(-\frac{\eta}{2t}\right), \qquad \frac{\partial c}{\partial r} = \frac{1}{\sqrt{t}}\frac{dF}{d\eta}, \qquad \frac{\partial^2 c}{\partial r^2} = \frac{1}{t}\frac{d^2F}{d\eta^2}\,. \tag{2.34}$$

This allows us to write the new form of heat equation:

$$-\left(\frac{\eta}{2t}\right)\frac{dF}{d\eta} = \frac{1}{t}\frac{d^2F}{d\eta^2} + \frac{2}{\eta\sqrt{t}}\frac{1}{\sqrt{t}}\frac{dF}{d\eta}\,. \tag{2.35}$$

Denoting $G = F_\eta$ leads to

$$\frac{\partial G}{\partial \eta} = -\left(\frac{\eta}{2} + \frac{2}{\eta}\right) G\,, \tag{2.36}$$

which may be integrated immediately

$$G = \frac{A}{\eta^2} e^{-\eta^2/4}\,, \tag{2.37}$$

where $G \to 0$ as $\eta \to \infty$. Integrating by parts then determines

$$F = A\left[\frac{\sqrt{\pi}}{2}\operatorname{erfc}\left(\frac{\eta}{2}\right) - \frac{e^{-\eta^2/4}}{\eta}\right], \tag{2.38}$$

where $F \to 0$ as $\eta \to \infty$. In terms of the original solute concentration,

$$c = A\left[\frac{\sqrt{\pi}}{2}\operatorname{erfc}\left(\frac{r}{2\sqrt{t}}\right) - \frac{\sqrt{t}}{r}e^{-r^2/4t}\right]. \tag{2.39}$$

The mass balance (2.31b) may be written as

$$\beta\frac{dr_p}{dt} = \left.\frac{\partial c}{\partial r}\right|_{r=r_p} = \left.\frac{1}{\sqrt{t}}\frac{dF}{d\eta}\right|_{\eta=r_p/\sqrt{t}}. \tag{2.40}$$

As with the Neumann solution, the time dependence is eliminated with the assumption $r_p = 2\lambda\sqrt{t}$. The boundary is then located at $\eta = r_p(t)/\sqrt{t} = 2\lambda$, and the above equation provides a relation for A in terms of β and λ:

$$\beta\lambda = \frac{A}{4\lambda^2}e^{-\lambda^2}. \tag{2.41}$$

At $r = r_p$, the concentration $c(r_p, t) = F(2\lambda) = -1$ is constant. Equation (2.38) then provides a transcendental equation for λ:

$$2\beta\lambda^2\left[1 - \sqrt{\pi}\lambda e^{\lambda^2}\operatorname{erfc}(\lambda)\right] = 1. \tag{2.42}$$

In Fig. 2.3, we show the concentration $c(r, t)$ for $r \geq r_p$, $\beta = 8$ at times $t = 0.1, 1, 5$. According to Eq. (2.42), this value of β results in $\lambda = 0.4065$. The starting point for each curve, $r_p = 2\lambda\sqrt{t} \approx 0.257, 0.813, 1.82$, shows how the sphere radius grows with time. The increase in the width of the concentration curve with time demonstrates how the concentrate depletes near the sphere boundary. As the depletion progresses, the gradient $c_r(r_p, t)$ decreases, resulting in a slower growth rate with time (as expected from the relation $dr_p/dt = \lambda/\sqrt{t}$).

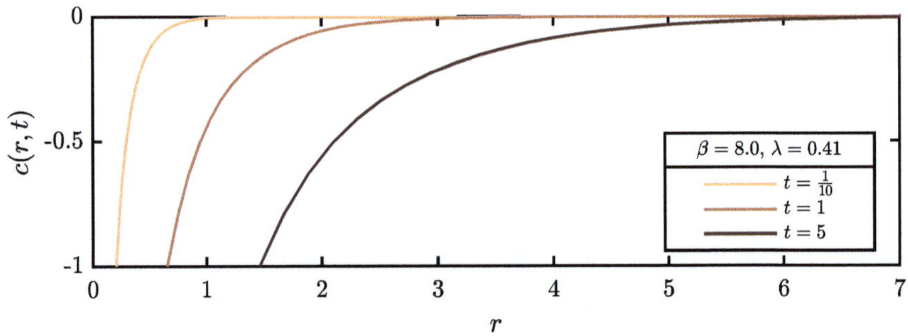

Fig. 2.3 Variation of concentration $c(r, t)$, Eq. (2.39), for one-phase spherical growth by diffusion for $\beta = 8$ (resulting in $\lambda = 0.4065$) at times $t = 0.1, 1, 5$. The corresponding sphere's radius is $r_p \approx 0.26, 0.81, 1.8$

2.1.5 Travelling Wave Forms

When the independent variables x and t do not appear explicitly in the governing equations, we can always look for a special form of similarity solution, namely, the travelling wave. A travelling wave is a specific form of wave moving with constant speed and retaining its shape throughout the process. Travelling waves are found in many physical situations, although possibly the most famous case is the soliton, which was first reported in 1844 [135]. As the name suggests, travelling waves satisfy the wave equation. They involve a particular form of boundary-fixing transformation, working in terms of the variable $\eta(x, t) = x - vt$ for some *a priori* unknown constant velocity v. The point $\eta = 0$ is chosen to define a certain position on the wave, for example, it may represent the front of a wave with compact support or a fixed height, such as the mid-point. Since the front speed $s_t = v$ is constant then $s = vt$ (provided, of course, that $s(0) = 0$). Clearly this is a strong restriction on the solution form.

In terms of the new variable, $T(x, t) = F(\eta)$, the derivatives of the heat equation become

$$\frac{\partial T}{\partial t} = \frac{dF}{d\eta}\frac{\partial \eta}{\partial t} = -v\frac{dF}{d\eta}, \tag{2.43}$$

$$\frac{\partial T}{\partial x} = \frac{dF}{d\eta}\frac{\partial \eta}{\partial x} = \frac{dF}{d\eta}, \qquad \frac{\partial^2 T}{\partial x^2} = \frac{d^2 F}{d\eta^2}, \tag{2.44}$$

and the one-phase heat equation (2.29a) may be written as

$$\frac{d^2 F}{d\eta^2} + v\frac{dF}{d\eta} = 0. \tag{2.45}$$

This is a constant coefficients ordinary differential equation (ODE), which is easily solved. Imposing the interface temperature $F(0) = 0$ and the Stefan condition, $\beta v = F_\eta(0)$ leads to

$$F(\eta) = \beta(1 - e^{-v\eta}), \tag{2.46}$$

or, in terms of the original variables,

$$T(x, t) = \beta(1 - e^{-v(x-vt)}). \tag{2.47}$$

This solution form was first presented by Stefan in 1889 [147].

So here we have an exact solution to a one-phase Stefan problem, where the phase change front moves with constant velocity $s_t = v$. However, it is worth noting that at $x = 0$, this constant value requires an exponentially increasing temperature:

$$T(0, t) = \beta(1 - e^{v^2 t}). \tag{2.48}$$

This unrealistic temperature variation is a common problem with similarity solutions: the boundary conditions are forced to match the solution form hence, they often turn out to be unphysical.

The reader may now be questioning why, after making such bold statements in the introduction of focussing on physically realistic situations, such an unrealistic solution has been presented. The reason being that travelling waves provide a very useful tool in the analysis of many practically useful models described by systems analogous to Stefan problems. One example concerns the removal of environmental contaminants using adsorption columns. Adsorption columns are employed in a vast range of contaminant removal processes such as in water treatment or the removal of greenhouse gases. The process involves passing a fluid through a tube filled with a material capable of capturing certain components that are present in the fluid. The most common form of model involves a one-dimensional advection-diffusion equation for the cross-sectionally averaged contaminant concentration including a mass sink, while the interface is defined as the position where the concentration is zero. This combination of an advection-diffusion equation with a kinetic condition leads to a system analogous to a one-phase Stefan problem with a moving liquid phase (this problem is discussed in the examples at the end of the chapter). A whole range of such problems may be found in [3, 107, 118, 120, 121]. A similar travelling wave approach has been employed to understand biological invasion processes (both cellular and ecological); see [45].

2.2 Approximate Solution Methods

2.2.1 *Boundary Fixing Transformations*

An obvious difficulty in dealing with Stefan problems is the fact that the domain is continually changing and, even worse, one of the regions does not initially exist. This can be particularly problematic when developing a numerical solution. There are a number of ways to bypass this issue, for example, by analysing the small time behaviour, which is usually quite distinct to that over a longer time-scale (we look into this in Sect. 2.2.3). An alternative approach is to introduce a boundary-fixing transformation [35]. In itself, this is not an approximate solution, or even a solution, and in fact, we have already used two boundary-fixing transformations as part of the similarity solution and travelling wave analysis. However, boundary-fixing transformations are important in developing approximate solutions and this subsection had to go somewhere!

In one dimension, the most obvious transformation is to work in terms of the variable suggested by Landau,

$$\eta = \frac{x}{s(t)}, \tag{2.49}$$

so that the interface is located at $\eta = 1$ for all time. Defining $T(x,t) = F(\eta(x,t),t), \theta(x,t) = G(\eta(x,t),t)$, then

$$\frac{\partial T}{\partial x} = \frac{\partial F}{\partial \eta}\frac{\partial \eta}{\partial x} = \frac{1}{s}\frac{\partial F}{\partial \eta}, \qquad \frac{\partial^2 T}{\partial x^2} = \frac{1}{s^2}\frac{\partial^2 F}{\partial \eta^2}, \tag{2.50}$$

$$\frac{\partial T}{\partial t} = \frac{\partial F}{\partial t} + \frac{\partial F}{\partial \eta}\frac{\partial \eta}{\partial t} = \frac{\partial F}{\partial t} - \frac{\eta}{s}\frac{ds}{dt}\frac{\partial F}{\partial \eta}, \tag{2.51}$$

and the heat equation in the solid may be written as

$$s^2\frac{\partial F}{\partial t} - s\eta\frac{ds}{dt}\frac{\partial F}{\partial \eta} = \frac{\partial^2 F}{\partial \eta^2}, \tag{2.52}$$

with a similar form in a stationary liquid. The Stefan condition becomes

$$\beta s\frac{ds}{dt} = \left(\frac{\partial F}{\partial \eta} - \frac{\partial G}{\partial \eta}\right)\bigg|_{\eta=1}. \tag{2.53}$$

The Neumann solution of Sect. 2.1.1 has already demonstrated that with a fixed boundary temperature $s \propto \sqrt{t}$, $s_t \propto 1/\sqrt{t}$. This behaviour could cause problems for a numerical scheme starting at $t = 0$. The problem may be avoided by defining a new function $z = s^2$ (see [92]), which transforms the above heat equation and Stefan condition to

$$z\frac{\partial F}{\partial t} - \frac{\eta}{2}\frac{dz}{dt}\frac{\partial F}{\partial \eta} = \frac{\partial^2 F}{\partial \eta^2}, \tag{2.54}$$

$$\frac{\beta}{2}\frac{dz}{dt} = \left(\frac{\partial F}{\partial \eta} - \frac{\partial G}{\partial \eta}\right)\bigg|_{\eta=1}. \tag{2.55}$$

The term ss_t of Eq. (2.52) has transformed to z_t in Eq. (2.54), and the equations are now in a form amenable to standard numerical techniques.

On the domain $x \in (s, \infty)$, a more appropriate transformation may be $\xi(x,t) = x - s(t)$ (as employed with the travelling waves), such that $T(x,t) \to F(\xi(x,t),t)$. The heat equation in this case transforms to

$$\frac{\partial F}{\partial t} - \frac{ds}{dt}\frac{\partial F}{\partial \xi} = \frac{\partial^2 F}{\partial \xi^2}. \tag{2.56}$$

This produces a travelling wave form when the functional dependence on t is dropped, so losing the first term on the left-hand side and requiring $s_t = v$. The

Stefan condition becomes

$$\beta\frac{ds}{dt} = \left(\frac{\partial F}{\partial \xi} - \frac{\partial G}{\partial \xi}\right)\bigg|_{\xi=0} . \tag{2.57}$$

Note well! Although it may appear that the boundary-fixing transformations were already dealt with during the Neumann solution and travelling wave sections, the forms presented here are more general. The Neumann solution involves a transformation where the temperature becomes a function with only a single variable $T(x,t) = F(\eta)$, the present more general form has $T(x,t) = F(\eta,t)$, and the Neumann form may then be seen as a special case. The transform $\xi = x - s(t)$ differs from the travelling wave; there is no requirement that the solution form is fixed nor that the velocity is constant. Consequently, when we shift to the new dependent variable, we write $T(x,t) = F(\xi(x,t),t)$ (as opposed to $F(\xi(x,t))$ for the travelling wave).

Hill [64] extends the Landau transformation $\eta = x/s(t)$ by changing the time variable to $\tau = s(t)$. This second transformation is possible since $s(t)$ is strictly increasing or decreasing. In cylindrical and spherical domains,

$$\text{Cylindrical}: \quad \eta(r,t) = \frac{\ln r}{\ln R(t)}, \qquad \tau = \ln R(t), \quad T(r,t) = F(\eta,\tau) \tag{2.58}$$

$$\text{Spherical}: \quad \eta(r,t) = \frac{1-r}{1-R(t)}, \qquad \tau = 1 - R(t), \quad T(r,t) = \frac{F(\eta,\tau)}{r}. \tag{2.59}$$

Ribera et al. [133] combine boundary-fixing and temperature transformations to compare the accuracy of heat balance and perturbation methods against numerical solutions for Cartesian, cylindrical, and spherical problems. Noting that even without transforming the independent variables, the change $T(r,t) = F(r,t)/r$ transforms the spherical heat equation into the Cartesian version:

$$\frac{\partial T}{\partial t} = \frac{1}{r^2}\frac{\partial}{\partial r}\left(r^2\frac{\partial T}{\partial r}\right) \quad \Rightarrow \quad \frac{\partial F}{\partial t} = \frac{\partial^2 F}{\partial r^2} . \tag{2.60}$$

This equivalence suggests that Cartesian and spherical transformations will perform equally well and calculations have demonstrated this to be the case. For example, Ribera et al. [133] found excellent agreement between Cartesian and spherical approximate and numerical solutions, while significant differences were observed between the cylindrical solutions.

2.2.2 Perturbation Methods

Perturbation theory involves a variety of iterative methods that result in approximate solutions to problems involving a small parameter. In fact, it has been shown on many occasions that these methods are so versatile that it is often worthwhile introducing an artificial small parameter into a difficult problem only to later set this parameter to one to recover the original problem [14]. Perturbation theory is probably the most popular approximate method for solving Stefan problems: the pseudo-steady solution mentioned in Sect. 1.3.1 is simply the dominant term of a perturbation series.

There are a number of texts explaining the intricacies of perturbation theory (or asymptotic methods, the two terms are often used interchangeably [66, P. 177]). Bender and Orszag [14] contains an excellent chapter with many examples to practice on. Hinch's book [65] stems from a lecture course and provides a wide-ranging introduction to the topic. Then there are many practical applied mathematics books that utilise perturbation methods on real-world problems; two key examples are [54, 66].

To decide whether we have a small (or large) parameter requires knowledge of typical physical values. In Table 1.1, we presented the thermophysical parameters for a variety of common materials. Using the values from this table shows that to achieve a Stefan number $\beta (= L_m/(c_s \Delta T)) > 10$ for water requires $\Delta T < 16\,\text{K}$, for gold $\Delta T < 50\,\text{K}$, and for tin $\Delta T < 25\,\text{K}$. These values are easily achieved, so the message here is that it often makes sense to immediately assume $\beta \gg 1$ (unless otherwise indicated) and proceed accordingly.

From now on, to keep the algebra simple, we will deal with the one-phase Stefan problem formulated in Sect. 2.1.3. Denoting $\varepsilon = 1/\beta \ll 1$, then the Stefan condition, Eq. (2.29b), may be written $s_t = \varepsilon T_x$. Since ε is small, this indicates that the interface position is approximately stationary, $s_t \approx 0$. Physically, this may be viewed as a very slow motion of the front. This becomes obvious if we interpret large $\beta = L_m/(c\Delta T)$ as an indication that ΔT is relatively small, i.e. the temperature driving the phase change is not much different to the phase change temperature. In this case, it is natural to expect a slow process.

Given that the prime reason for solving a Stefan problem is usually to determine the position of the front, the fact that it appears stationary on the present time-scale is rather unsatisfactory. Hence, the analysis for the large Stefan number approximation begins by changing to a slow time variable τ, where $t = \tau/\varepsilon$. This results in a Stefan condition where the terms balance

$$\frac{ds}{d\tau} = \left.\frac{\partial T}{\partial x}\right|_{x=s}, \tag{2.61}$$

while the heat equation becomes

$$\varepsilon \frac{\partial T}{\partial \tau} = \frac{\partial^2 T}{\partial x^2}. \tag{2.62}$$

The presence of the small parameter suggests looking for a series solution:

$$T = T_0 + \varepsilon T_1 + \varepsilon^2 T_2 + \mathscr{O}(\varepsilon^3) , \tag{2.63}$$

where $\mathscr{O}(.)$ denotes *order*, and here we neglect terms smaller than order ε^2. This is the perturbation expansion where each term $T_i(x, \tau)$ is assumed to be of order unity. It is also an option to expand s. To demonstrate the different routes in this section, we will proceed with the simpler choice, and in the following section, we will provide an example where s is expanded.

Now, we substitute the series into Eq. (2.62), with the new time variable, $\tau = \varepsilon t$,

$$\varepsilon \left(\frac{\partial T_0}{\partial \tau} + \varepsilon \frac{\partial T_1}{\partial \tau} + \cdots \right) = \frac{\partial^2 T_0}{\partial x^2} + \varepsilon \frac{\partial^2 T_1}{\partial x^2} + \varepsilon^2 \frac{\partial^2 T_2}{\partial x^2} + \cdots . \tag{2.64}$$

Collecting terms of the same power of ε gives a sequence of simpler problems:

$$\frac{\partial^2 T_0}{\partial x^2} = 0 , \qquad \frac{\partial^2 T_1}{\partial x^2} = \frac{\partial T_0}{\partial \tau} , \qquad \frac{\partial^2 T_2}{\partial x^2} = \frac{\partial T_1}{\partial \tau} . \tag{2.65}$$

The boundary conditions (2.29c) at each order are

$$T_0(0, \tau) = -1 , \quad T_1(0, \tau) = T_2(0, \tau) = 0 , \tag{2.66}$$

$$T_0(s, \tau) = T_1(s, \tau) = T_2(s, \tau) = 0 . \tag{2.67}$$

Solving the first and second equations in (2.65) and applying the boundary conditions gives the leading and first-order solutions as

$$T_0 = \frac{x}{s} - 1 , \qquad T_1 = \frac{x}{6} \left(1 - \frac{x^2}{s^2} \right) \frac{ds}{d\tau} . \tag{2.68}$$

At leading order, the temperature is linear, as suggested by the form of the solid temperatures seen in Figs. 2.1 and 2.2.

We are now in a position to calculate the temperature derivatives and substitute into the Stefan condition to determine $s(\tau)$, but before doing this, we note that the above solution is as accurate as we can get in this formulation. If we try to continue to second order, we see from Eq. (2.65) that T_2 depends on the time derivative of T_1 and consequently on $s_{\tau\tau}$. Since s is a function of time alone, we may integrate the T_2 equation, but when we substitute into the Stefan condition, we obtain a second-order equation for s. With only a single initial condition, we will be stuck. This problem may be circumvented by either expanding s in powers of ε or by making another change of variables. Here, we choose the latter option and leave the former as an exercise for the enjoyment of the reader. First, we use the boundary-fixing transformation $\eta = x - s(\tau)$ that fixes the moving front to $\eta = 0$. Then, since s is monotonic in time, as suggested in the previous section, we may use it as a time variable so defining $u(\eta(x,\tau), s(\tau)) = T(x,\tau)$. The derivatives become

$$\frac{\partial T}{\partial \tau} = \frac{\partial u}{\partial \eta}\frac{\partial \eta}{\partial \tau} + \frac{\partial u}{\partial s}\frac{ds}{d\tau} = \frac{ds}{d\tau}\left(\frac{\partial u}{\partial s} - \frac{\partial u}{\partial \eta}\right) \tag{2.69}$$

$$\frac{\partial T}{\partial x} = \frac{\partial u}{\partial \eta}\frac{\partial \eta}{\partial x} = \frac{\partial u}{\partial \eta}, \qquad \frac{\partial^2 T}{\partial x^2} = \frac{\partial^2 u}{\partial \eta^2}. \tag{2.70}$$

The Stefan condition becomes

$$\frac{ds}{d\tau} = \left.\frac{\partial u}{\partial \eta}\right|_{\eta=0} \tag{2.71}$$

and hence the new form of heat equation:

$$\frac{\partial^2 u}{\partial \eta^2} = \varepsilon\frac{ds}{d\tau}\left(\frac{\partial u}{\partial s} - \frac{\partial u}{\partial \eta}\right) = \varepsilon \left.\frac{\partial u}{\partial \eta}\right|_{\eta=0}\left(\frac{\partial u}{\partial s} - \frac{\partial u}{\partial \eta}\right). \tag{2.72}$$

This is subject to the boundary conditions:

$$T(0,t) = u(-s,s) = -1, \quad T(s,t) = u(0,s) = 0. \tag{2.73}$$

Taking a perturbation series for $u(\eta, s) = u_0 + \varepsilon u_1 + \cdots$ leads to the following sequence of equations:

$$\frac{\partial^2 u_0}{\partial \eta^2} = 0, \tag{2.74}$$

$$\frac{\partial^2 u_1}{\partial \eta^2} = \left.\frac{\partial u_0}{\partial \eta}\right|_{\eta=0}\left(\frac{\partial u_0}{\partial s} - \frac{\partial u_0}{\partial \eta}\right), \tag{2.75}$$

$$\frac{\partial^2 u_2}{\partial \eta^2} = \left.\frac{\partial u_0}{\partial \eta}\right|_{\eta=0}\left(\frac{\partial u_1}{\partial s} - \frac{\partial u_1}{\partial \eta}\right) + \left.\frac{\partial u_1}{\partial \eta}\right|_{\eta=0}\left(\frac{\partial u_0}{\partial s} - \frac{\partial u_0}{\partial \eta}\right). \tag{2.76}$$

After applying the appropriate boundary conditions, we obtain

$$u_0 = \frac{\eta}{s}, \qquad u_1 = -\left[\frac{\eta^3}{6s^3} + \frac{\eta^2}{2s^2} + \frac{\eta}{3s}\right] = -\frac{\eta}{6s^3}(\eta + s)(\eta + 2s). \tag{2.77}$$

It is easily verified that these coincide with the previous solutions (2.68) after replacing $\eta = x - s$ and noting that at leading order $s_\tau = (\partial u_0/\partial \eta)|_{\eta=0} = 1/s$. The key difference in the two formulations is that the time derivative does not appear on the right-hand side of (2.72), and as a result, there is no sequence of increasing order time derivatives in the equations for u_i. Consequently, we may now substitute for u_0, u_1 into Eq. (2.76) and integrate to find

$$u_2 = \frac{1}{40}\frac{\eta^5}{s^5} + \frac{1}{8}\frac{\eta^4}{s^4} + \frac{5}{18}\frac{\eta^3}{s^3} + \frac{1}{3}\frac{\eta^2}{s^2} + \frac{7}{45}\frac{\eta}{s} . \tag{2.78}$$

If desired, or if we have the energy, we may continue the process solving for any u_i with increasingly long, but simple to calculate, algebraic expressions. Substituting for u_i into the Stefan condition, Eq. (2.71), leads to

$$s\frac{ds}{d\tau} = \left(1 - \frac{1}{3}\varepsilon + \frac{7}{45}\varepsilon^2 + \cdots\right) , \tag{2.79}$$

and hence

$$s = \sqrt{2\tau}\left(1 - \frac{1}{6\beta} + \frac{23}{360\beta^2} + \cdots\right) . \tag{2.80}$$

A check on any perturbation solution is that in the limit $\varepsilon \to 0$, it must coincide with the expansion of the exact solution (if an exact solution is available). Going back to the one-phase solution, Eqs. (2.30), we have $s = 2\lambda\sqrt{t}$ where

$$\sqrt{\pi}\lambda \operatorname{erf}(\lambda)e^{\lambda^2} = \frac{1}{\beta} = \varepsilon . \tag{2.81}$$

In the limit $\beta \to \infty$ ($\varepsilon \to 0$), this only balances if $\lambda \to 0$. The small λ expansion leads to

$$2\lambda\left(\lambda - \frac{\lambda^3}{3} + \cdots\right)\left(1 + \lambda^2 + \cdots\right) = 2\lambda^2\left(1 + \frac{2\lambda^2}{3} + \cdots\right) = \varepsilon . \tag{2.82}$$

So at leading order $\lambda = \sqrt{\varepsilon/2}$, which suggests the more complete expansion form:

$$\lambda = \sqrt{\frac{\varepsilon}{2}}\left(1 + \varepsilon\lambda_1 + \varepsilon^2\lambda_2 \cdots\right) . \tag{2.83}$$

Substituting this series back into Eq. (2.82) then, after some manipulation, we obtain $\lambda_1 = -1/6, \lambda_2 = 23/360$, thus confirming equation (2.80). So this solution is obviously accurate at least to second order in $\varepsilon = 1/\beta$.

2.2.3 Small and Large Time Solutions

Small and large time solutions allow us to study the limiting behaviour of the system, as the name says, when $t \to 0$ or $t \to \infty$. Small time solutions play a particularly important role in understanding how the boundary conditions first affect the phase change process (since the phase change front is initially at the boundary).

They are also very useful in the development of numerical solutions, particularly when the boundary conditions force a singularity on the solution. If we can find a small time solution, then we may use this as the initial condition for a numerical scheme and so avoid the singularity. For example, starting at some arbitrary time $t = t_0 \ll 1$, we replace the conditions $T(x, 0)$, $s(0)$ with $T(x, t_0)$, $s(t_0)$, which come from the small time solution.

In fact, small and large time is relative: in the previous chapter, we first nondimensionalised time to balance both sides of the heat equation, but this resulted in a small parameter $\varepsilon = 1/\beta$ on the right-hand side of the Stefan condition (1.21). This was quite distressing since it meant the very effect under study, namely, the phase change, was negligible. For this reason, a second time-scale was introduced, which balanced terms in the Stefan condition but introduced a small parameter into the heat equation. So, when we now talk about a small time-scale, this could simply mean reverting to the thermal time-scale or some other time-scale that changes the balance in the equations to highlight the dominant effects in the very early stages of the process. Similarly with large time-scales, the phase change time-scale could be thought of as a large time scaling of the thermal problem, etc.

In Sect. 3.1, we introduce the flow of a thin liquid layer. If we were to work on the flow time-scale, which is much shorter than that of solidification, then the dominant flow problem would simply be that predicted by standard lubrication theory on a stationary substrate. In Mitchell et al. [92], small time solutions are examined for the sole purpose of identifying suitable initial conditions for finite element schemes.

Consider the one-phase Stefan problem defined by

$$\frac{\partial \theta}{\partial t} = \frac{\partial^2 \theta}{\partial x^2}, \qquad s < x < \infty \tag{2.84}$$

$$\beta \frac{ds}{dt} = -\left.\frac{\partial \theta}{\partial x}\right|_{x=s}. \tag{2.85}$$

These are to be solved subject to

$$\theta(s, t) = 0, \qquad \theta|_{x\to\infty} = \theta(x, 0) = -1. \tag{2.86}$$

This system describes the solidification of a supercooled liquid originally at temperature $\theta(x, 0) = -1$. Solidification proceeds from the boundary $x = 0$. For the one-phase problem, we assume that the solid remains at the phase change temperature (solidus).

The term supercooled refers to a liquid that is below its freezing point without solidifying. Supercooled liquids typically solidify much more rapidly than a non-supercooled liquid. At sufficiently high solidification rates, the liquid may not have time to form its usual crystalline structure. Materials made from supercooled melts can therefore have markedly different properties to the standard form of the material; they can present greater corrosion resistance, toughness, strength, hardness, and elasticity than common materials. Metal alloys formed in this way can be twice as strong and three times more elastic than steel!

In the absence of nucleation sites, water may be cooled down to around −40 °C [94]. This causes great difficulty for aircraft flying through clouds when the ambient temperature is below freezing. The droplets in the cloud are supercooled and then a fast-moving aircraft appears, presenting a plethora of nucleation sites. For this reason, the outer surface of aircraft is usually heated when flying in such conditions: a typical system for a long-range passenger jet consumes of the order 10^5W when active [114].

We now define the small parameter $\varepsilon = 1/\beta$ and look for a perturbation solution with

$$\theta = \theta_0 + \varepsilon\theta_1 + \cdots \qquad s = s_0 + \varepsilon s_1 + \cdots . \tag{2.87}$$

The presence of the small parameter in the equation for s, Eq. (2.85), suggests that we should treat it with more care than when it appeared in the heat equation. Working to first order, the system is governed by

$$\frac{\partial}{\partial t}(\theta_0 + \varepsilon\theta_1) = \frac{\partial^2}{\partial x^2}(\theta_0 + \varepsilon\theta_1), \quad s < x < \infty \tag{2.88}$$

$$\frac{d}{dt}(s_0 + \varepsilon s_1) = -\varepsilon\left(\frac{\partial}{\partial x}(\theta_0 + \varepsilon\theta_1)\bigg|_{x=s_0+\varepsilon s_1}\right) \tag{2.89}$$

$$= -\varepsilon\,\frac{\partial\theta_0}{\partial x}\bigg|_{x=s_0} + \mathscr{O}(\varepsilon^2). \tag{2.90}$$

The right-hand side of the final expression comes through taking the Taylor series of θ around $x = s_0$. The boundary conditions (to first order) become

$$\begin{aligned}\theta_0(s_0 + \varepsilon s_1, t) &+ \varepsilon\theta_1(s_0 + \varepsilon s_1, t)\\ &= \theta_0(s_0, t) + \varepsilon s_1\,\frac{\partial\theta_0}{\partial x}\bigg|_{x=s_0} + \varepsilon\theta_1(s_0, t) = 0,\end{aligned} \tag{2.91}$$

$$\theta_0|_{x\to\infty} + \varepsilon\,\theta_1|_{x\to\infty} = -1, \qquad \theta_0(x, 0) + \varepsilon\theta_1(x, 0) = -1, \tag{2.92}$$

where again a Taylor expansion has been used at $x = s$.

The leading order Stefan condition, subject to $s_0(0) = 0$, determines $s_0(t) = 0$, and the boundary appears fixed at the initial position. The leading order temperature therefore satisfies a fixed boundary problem:

$$\frac{\partial\theta_0}{\partial t} = \frac{\partial^2\theta_0}{\partial x^2}, \qquad 0 < x < \infty, \tag{2.93}$$

$$\theta_0(0, t) = 0, \quad \theta_0|_{x\to\infty} = \theta_0(x, 0) = -1. \tag{2.94}$$

The solution may be obtained via a similarity reduction or Laplace transform:

$$\theta_0 = -1 + \text{erfc}\frac{x}{2\sqrt{t}} . \tag{2.95}$$

This permits the calculation of s_1. The first-order Stefan condition states that

$$\frac{\partial s_1}{\partial t} = -\left.\frac{\partial \theta_0}{\partial x}\right|_0 = \frac{1}{\sqrt{\pi t}}, \quad \Rightarrow \quad s_1 = 2\sqrt{\frac{t}{\pi}} . \tag{2.96}$$

The temperature θ_0 and interface position define the leading order small time solution for the solidification process. The series may be continued to determine expressions for θ_n for $n > 1$ to find a more accurate approximation. However, if the goal is to find a suitable initial condition for a numerical scheme of the full problem, then it is sufficient to choose a sufficiently small $t = t_0 \ll 1$ and apply the initial conditions defined by θ_0, s_1:

$$\theta(x, t_0) = -1 + \text{erfc}\frac{x}{2\sqrt{t_0}}, \qquad s(t_0) = \frac{2}{\beta}\sqrt{\frac{t_0}{\pi}} . \tag{2.97}$$

2.2.4 Heat Balance Integral Method (HBIM)

The Heat Balance Integral Method involves defining an approximating function for the temperature, the classic one being a quadratic equation, and then introducing this to an integrated heat equation and boundary conditions to determine the unknown coefficients. It is a technique that has been largely neglected by the mathematical community, possibly due to the rather *ad hoc* choice of approximating function. This may be appreciated by consulting Hill's book [64] where the majority of chapters involve a range of heat balance techniques. The review of Mitchell and Myers [89] presents seven different formulations for the same problem, even when employing just a quadratic approximating function. Some of the guesswork is removed by adopting an *opTimal Integral Method* (TIM) that determines the most accurate approximation in a least-squares sense [98, 100]. The *Combined Integral Method* (CIM) of [112] may be preferable for time-dependent boundary conditions, although slightly less accurate for standard problems. In [112], it is shown that for physically realistic Stefan numbers, the CIM (and hence the TIM) is more accurate than the second-order perturbation for a standard one-phase problem. When the perturbation is more accurate, the least-squares errors are so small (below 0.1%) that it is irrelevant which approximation method is employed. The TIM and CIM have the added advantage that there is no need for a small parameter. They both significantly outperform perturbation methods when β is small. Despite the advances in applying heat balance methods, they still generally appear in the engineering literature, while perturbation solutions are predominant in the mathematical literature.

To explain heat balance methods, we will start with the original form of Goodman [56] applied to a thermal problem without phase change. Subsequently, we will demonstrate the TIM and other variations. Goodman [56] originally considered the following problem:

$$\frac{\partial T}{\partial t} = \frac{\partial^2 T}{\partial x^2}\,, \qquad 0 < x < \infty \tag{2.98a}$$

$$T(0,t) = -1\,, \qquad T|_{x\to\infty} \to 0\,, \qquad T(x,0) = 0\,. \tag{2.98b}$$

The system defined by Eqs. (2.98a, 2.98b) has the exact solution $T = \mathrm{erfc}(x/(2\sqrt{t})$ (obtained by defining a similarity solution or using a Laplace transform).

To apply the original HBIM requires three simple steps:

1. Introduce the heat penetration depth, $\delta(t)$. For $x \geq \delta$, the temperature rise above the initial value is negligible.
2. Define an approximate function for the temperature, typically a polynomial, and apply sufficient boundary conditions at $x = 0$ and $\delta(t)$ to determine all unknown coefficients in terms of the unknown function δ.
3. Integrate the governing equation for $x \in [0, \delta]$ to produce what is termed the heat balance integral. Typically, this results in an ordinary differential equation for δ.

The more discerning reader may observe that this is simply an adaptation of the earlier Karman-Pohlhausen integral method for analyzing boundary layers [90].

The standard HBIM developed by Goodman [56] involves a quadratic approximating function:

$$T = a_0 + a_1\left(1 - \frac{x}{\delta}\right) + a_2\left(1 - \frac{x}{\delta}\right)^2 . \tag{2.99}$$

There are four unknowns here, a_i and δ. The two boundary conditions plus the heat balance integral are not sufficient to determine them all. For this reason, a smoothness condition is also imposed, namely, that the temperature must approach the far-field constant value smoothly, this requires $T_x(\delta, t) = 0$. Since δ varies with time, we must impose an initial condition; at $t = 0$, the temperature is constant $T(x, 0) = 0$ indicating no heat penetration and hence $\delta(0) = 0$. Applying the boundary conditions, $T(0, t) = -1$, $T(\delta, t) = T_x(\delta, t) = 0$ leads to

$$T = -\left(1 - \frac{x}{\delta}\right)^2 , \qquad 0 < x < \delta\ . \tag{2.100}$$

The heat balance integral simply involves integrating the heat equation:

$$\int_0^\delta \frac{\partial T}{\partial t}\,dx = \int_0^\delta \frac{\partial^2 T}{\partial x^2}\,dx\ . \tag{2.101}$$

The right-hand side integrates immediately. For the left, we invoke Leibniz' theorem:

$$\frac{\partial}{\partial t}\int_{a(t)}^{b(t)} f(x,t)\,dx = \int_{a(t)}^{b(t)} \frac{\partial f}{\partial t}\,dx + \frac{db}{dt}f(b,t) - \frac{da}{dt}f(a,t)\,. \quad (2.102)$$

Noting that the lower limit of the integrals in (2.101) is constant (zero in fact) and also that $T(\delta, t) = 0$, we find that

$$\int_0^{\delta} \frac{\partial T}{\partial t}\,dx = \frac{\partial}{\partial t}\int_0^{\delta} T\,dx\,. \quad (2.103)$$

Since $T_x(\delta, t) = 0$, the integrated heat equation reduces to

$$\frac{\partial}{\partial t}\int_0^{\delta} T\,dx = -\left.\frac{\partial T}{\partial x}\right|_{x=0}\,. \quad (2.104)$$

Substituting for the temperature via Eq. (2.100) and evaluating the integral results in an ODE for δ

$$\frac{\partial}{\partial t}\int_0^{\delta}\left[-\left(1-\frac{x}{\delta}\right)^2\right]dx = \frac{\partial}{\partial t}\left(\frac{\delta}{3}\right) = \frac{2}{\delta}\,. \quad (2.105)$$

Applying $\delta(0) = 0$ determines $\delta = \sqrt{12t}$, and the temperature profile takes the form

$$T = -\left(1 - \frac{x}{\sqrt{12t}}\right)^2, \qquad 0 < x < \sqrt{12t}\,. \quad (2.106)$$

It is worth pointing out that although, as the name suggests, the HBIM was developed for thermal problems, it has a variety of other applications. Its greatest impact seems to be with Stefan problems, since very few analytical solutions exist. However, the Schrödinger equation may be considered as a heat equation with a complex diffusivity. The heat equation is used in probability and describes random walks. For this reason, it is also applied in financial mathematics and is a particular limit of the famous Black-Scholes equation. It is important in Riemannian geometry and thus topology. In viscous flow, the heat equation is retrieved in the analysis of an impulsively moved plate in a semi-infinite viscous fluid. Consequently, there are many analogous problems available for a heat balance type analysis.

Since the HBIM is an adaptation of the Karman-Pohlhausen integral method for analysing boundary layers in fluid flow, it should come as no surprise that the HBIM has been applied to analyse boundary layers. Papers such as [80, 99] exploit this link to investigate boundary layer flow of power law fluids and nanofluids. The method has even been applied to the Korteweg-de Vries equation [111].

2.2.4.1 Improved Heat Balance Formulations

The original HBIM was based on a quadratic approximating function however in the search for greater accuracy authors first increased the degree of the fitting polynomial, adding more unknowns, then included non-integer powers, exponential and piece-wise linear approximating functions were introduced. Sometimes the new forms improved the accuracy, sometimes they simply obfuscated the method, and some were simply generalisations of known exact solutions, making the approximation irrelevant; details and different formats are discussed in the review [89]. In all cases, the choice was motivated by knowledge of an exact or numerical solution. A further alternative, the Refined Integral Method (RIM), was introduced in [136]. The RIM involves integrating the heat equation twice and is in fact identical to the "Integral formulation by integration" described much earlier in Hill's book [64]. One ambiguity with the RIM is that the order in which the double integration is evaluated results in distinct formulations and hence different accuracies. The Combined Integral Method (CIM) introduced in [112] merges HBIM and RIM formulations and removes this particular ambiguity. The opTimal Integral Method (TIM) introduces a further unknown, the highest order exponent, which is then determined through a least-squares error [98, 100]. Generally, the CIM is less accurate than the TIM; however, it has proved valuable in cases where the optimal exponent appears to be time dependent, for example, when the boundary condition is time dependent. The relative accuracy of the HBIM, RIM, and an alternative RIM is examined in [90] for various approximating functions. Methods of defining the optimal exponent (in a least-squares sense), through local or global matching, are investigated in [91]. In the following, we will focus on the TIM.

Frustrated by the random choice of approximating function or a choice based on an already known solution (hence removing the need for the approximation), Myers [98, 100] developed the opTimal Integral Method (TIM). The TIM involves an approximating function where the exponent is initially unknown. If we consider the thermal problem defined by Eqs. (2.98), then we may write the approximating function as

$$T(x,t) = a_0 + a_1\left(1 - \frac{x}{\delta}\right) + a_n\left(1 - \frac{x}{\delta}\right)^n , \tag{2.107}$$

where n is initially unknown. After applying the boundary conditions, this reduces to a generalised form of (2.100):

$$T(x,t) = -\left(1 - \frac{x}{\delta}\right)^n . \tag{2.108}$$

A new unknown requires an additional equation. In [98, 100], the least-squares error is introduced:

$$E_n(t) = \int_0^\delta \left[\frac{\partial T}{\partial t} - \frac{\partial^2 T}{\partial x^2} \right]^2 dx \geq 0 \,. \tag{2.109}$$

If T is an exact solution, then obviously $E = 0$. Approximate solutions will have $E > 0$. Taking the square of $T_t - T_{xx}$ prevents the cancelling of errors of opposite sign and also magnifies the importance of regions where T does not closely satisfy the heat equation. The value of n is chosen to minimise this error.

Applying the heat balance integral to the temperature profile (2.108) leads to $\delta = \sqrt{2n(n+1)t}$ (for $n = 2$ Goodman's version, Eq. (2.106) with $\delta = \sqrt{12t}$, is retrieved). Then, substituting for T into Eq. (2.109), we obtain an analytical expression for the error:

$$E_n = \frac{(6n^2 + 2n^4 + 2n - 7n^3 - 1)\sqrt{2n(n+1)}}{4(n+1)^2(2n-3)(4n^2-1)} t^{-3/2} = e_n t^{-3/2} \,. \tag{2.110}$$

The problem now reduces to minimising e_n. The range of n is restricted only by the requirement of a smooth transition to the far-field solution that determines $n \geq 2$ (so the singularities in e_n are out of the range of interest). For the fixed temperature boundary condition, this occurs when $n \approx 2.233$. The fact that this value is so close to $n = 2$ indicates why the original quadratic solution is so popular (and accurate) for fixed temperature boundary conditions. This is rather fortuitous, but it is not always the case.

Consider now the same problem but with a Neumann condition $T_x(0, t) = -1$ at the boundary. This has the exact solution:

$$T = 2\sqrt{\frac{t}{\pi}} e^{-x^2/(4t)} - x \,\mathrm{erfc}\left(\frac{x}{2\sqrt{t}} \right) . \tag{2.111}$$

Subject to the new condition, the TIM requires

$$T = \frac{\delta}{n} \left(1 - \frac{x}{\delta} \right)^n . \tag{2.112}$$

The heat balance integral then leads to

$$\delta = \sqrt{n(n+1)t} \,. \tag{2.113}$$

In this case, the least-squares error calculation determines $n \approx 3.584$. In fact, this same value minimises the error for both Neumann and Robin conditions; see [98]. This is far from the classical value $n = 2$. The difference is demonstrated in Fig. 2.4 where the exact solution (2.111), the TIM (Eq. (2.112) with $n = 3.584$), and HBIM solutions (Eq. (2.112) with $n = 2$) are shown at time $t = 1$. The

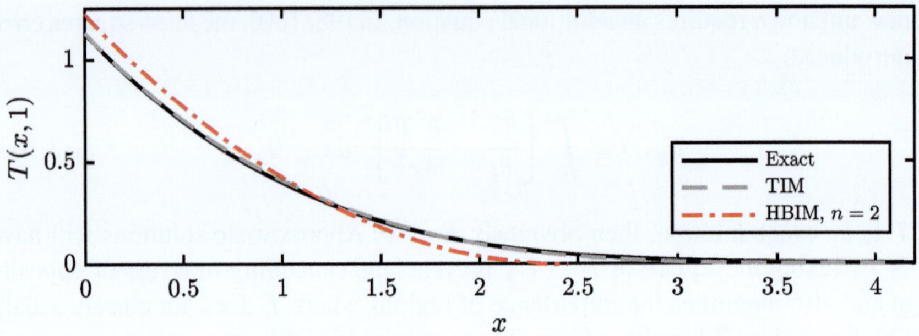

Fig. 2.4 Comparison of temperature profiles for a semi-infinite media subject to a constant flux boundary condition $T_x(0, t) = -1$ at time $t = 1$ using the exact solution (2.111) (solid line), TIM solution (Eq. (2.112) with $n = 3.584$) (dashed), and classical HBIM (Eq. (2.112) with $n = 2$) (dash-dot)

solid and dashed lines represent the exact and TIM solutions, respectively: they are virtually indistinguishable. The dash-dot line is the HBIM profile that is clearly a poor approximation.

For thermal problems, the TIM can significantly improve accuracy over the standard HBIM formulation, in particular for Neumann and Robin boundary conditions. It also has the distinct advantage of being simple to use. For a fixed temperature/Dirichlet boundary condition, the temperature is given by Eq. (2.108) with $n = 2.233$ and for a Neumann condition by (2.112) with $n = 3.584$. For the Robin condition, $T_x(0, t) = T(0, t) - 1$, by

$$T = \frac{\delta}{n+\delta}\left(1 - \frac{x}{\delta}\right)^n, \quad t = \frac{1}{n(n+1)}\left(\frac{\delta^2}{2} + n\delta - n^2 \ln\left(\frac{n+\delta}{n}\right)\right) \tag{2.114}$$

again with $n = 3.584$. This final form has an implicit relation for $\delta(t)$, which must be solved for a given time; see [98].

2.2.4.2 Application to Stefan Problems

The HBIM, RIM, CIM, and TIM can all provide simple solutions to standard thermal problems. However, their greatest impact is in the field of Stefan problems, where very few analytical solutions are known.

Focussing on the TIM for Stefan problems, a similar route to the above may be followed. If we consider the one-phase Stefan problem discussed in Sect. 2.1.3, then an appropriate temperature profile is

$$T(x, t) = a_0 + a_1\left(1 - \frac{x}{s}\right) + a_n\left(1 - \frac{x}{s}\right)^n . \tag{2.115}$$

In the case of melting, we do not require a zero gradient in the far field; hence, the restriction $n \geq 2$ applied in the previous thermal problems does not apply here. Applying $T(s,t) = 0$, $T(0,t) = -1$ determines $a_0 = 0$, $a_n = -(1 + a_1)$. The heat balance integral is now applied over the domain $[0, s(t)]$:

$$\frac{d}{dt} \int_0^s T\, dx = \int_0^s \frac{\partial^2 T}{\partial x^2}\, dx = T_x(s,t) - T_x(0,t)\,. \tag{2.116}$$

As discussed previously, there are various formulations for each integral method. Here, when evaluating $T_x(s,t)$, we may take the derivative of (2.115) and then substitute $x = s$; however, in Mitchell and Myers [89], the most accurate approximation was obtained by replacing $T_x(s,t) = \beta s_t$. Following this latter route and assuming a_n, n are constant, evaluating (2.116) and rearranging leads to

$$s\frac{ds}{dt} = \frac{2(n+1)((1+a_1)n - a_1)}{2 - a_1(n-1) + 2(n+1)\beta}\,. \tag{2.117}$$

The Stefan condition is

$$\beta\frac{ds}{dt} = \left.\frac{\partial T}{\partial x}\right|_{x=s} = -\frac{a_1}{s}\,. \tag{2.118}$$

Integrating and applying $s(0) = 0$ determines $s = \sqrt{-2a_1 t/\beta}$, indicating $a_1 < 0$ (since $s(t)$ is a real, non-negative quantity). Combining the Stefan condition and (2.117) provides a quadratic for a_1 in terms of n, β:

$$\frac{2(n+1)((1+a_1)n - a_1)}{2 - a_1(n-1) + 2(n+1)\beta} = -\frac{a_1}{\beta}\,, \tag{2.119}$$

where the negative root must be chosen. Finally, the value of n is determined through the least-squares error integral, equivalent to (2.109), where the integral limits are now $x = 0, s$

$$E_n(t) = \int_0^s \left[\frac{\partial T}{\partial t} - \frac{\partial^2 T}{\partial x^2}\right]^2 dx \geq 0\,. \tag{2.120}$$

In Myers [100], it is shown that the error is a minimum for $n \approx 1.8$, with a negligible dependence on β and $a_1 \approx -0.793$. *This is the optimal value when studying phase change due to a Dirichlet boundary condition.* Again this is close to $n = 2$, so we anticipate that the classical method (which assumes $n = 2$) will appear accurate for this situation.

This one-phase problem was previously tackled in Sect. 2.2.2, using a perturbation method, where we found both first- and second-order approximations. In Fig. 2.5, we present a comparison of results for $\beta = 1$ obtained using the exact solution, the TIM, and first- and second-order perturbations. The solid line

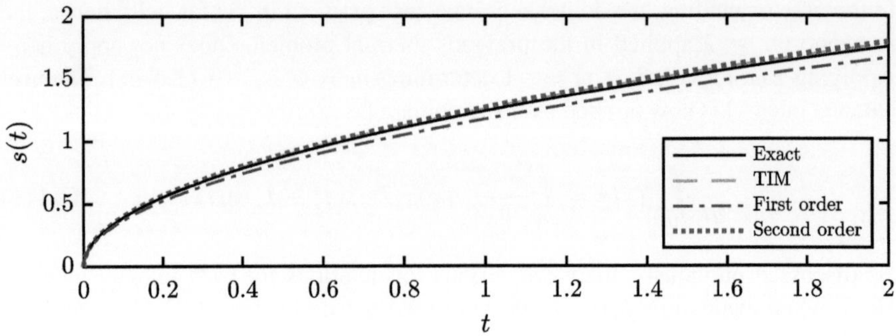

Fig. 2.5 Comparison of one-phase solutions using (**a**) exact solution (solid line); (**b**) TIM with $n = 1.8$, dashed line; (**c**) first-order perturbation (dash-dot); and (**d**) second-order perturbation (dotted). The value $\beta = 1$ is chosen to emphasise the difference in results

represents the exact solution $s = 2\lambda\sqrt{t}$, and λ satisfies equation (2.30). The perturbation solution has the same expression for s with λ determined by the appropriate order of Eq. (2.83). The TIM requires solving the quadratic equation (2.119) and minimising (2.120), this leads to $n \approx 1.8, a_n \approx -0.793$. The value of $\beta = 1$ has been chosen to emphasise the difference in solutions. This may seem a little unfair to the perturbation solution, which strictly holds for large $\beta = 1/\varepsilon$, but for sufficiently large β, all results become indistinguishable. The first-order perturbation, i.e. up to λ_1, is the dash-dot line – this is the worst approximation. The second-order solution is the dotted line that sits just above the TIM. The TIM is obviously the best approximation; however, if β is increased to 1.5, the second-order perturbation becomes the most accurate. When $\beta = 5$, all solution forms give excellent agreement. As β decreases below unity, the perturbation solutions rapidly reduce in accuracy. When $\beta = 0.16$, the first-order perturbation of Sect. 2.2.2 is negative, while the TIM is virtually indistinguishable from the exact solution. This demonstrates the versatility of the TIM, which is easily applied to large, small, and order one Stefan problems. A small Stefan number perturbation solution does exist but is notoriously difficult. It requires working on two time-scales $t \sim \beta^2$ and $t \sim \ln(1/\beta)$. To make analytical progress requires further approximations, which then compromises the matching; see [46]. Practically, the perturbation is best applied for large β values.

An important feature of the optimisation of the Stefan problem is that there are in fact a variety of formulations even with $n = 2$. Wood [156] discusses six, due to permutations such as taking $s - x$ instead of $(1 - x/s)$ or the use of $T_x(s, t)$. Above we use the Stefan condition to determine $T_x(s, t)$, rather than calculating the value from the chosen form of $T(x, t)$. This was motivated by the knowledge that it is the most accurate form; see [100]. Wood prefers $1 - x/s$ since it leads to a constant value for a_n. The different methods are discussed in Mitchell and Myers [90], where a seventh option is shown to exist; they also point out that a_1 is only constant for the Dirichlet condition; for Neumann and Robin conditions, it is time dependent, and the system reduces to two ordinary differential equations for a_n, s. While not the simple analytical form presented above for the Dirichlet condition, the reduction of a coupled PDE/ODE system, over a moving domain to two coupled ODEs, is still a significant simplification.

In Myers [100], a number of other physical problems are examined via TIM and T-RIM (twice integrated TIM) approaches, including ablation, the travelling wave, and a time-dependent boundary flux. The travelling wave is presented as a cautionary tale; as time increases, the heat balance solution deteriorates, suggesting heat balance methods for this case should only be applied for small times. Certain cases may lead to a time-dependent n; this issue is discussed in [112] where the Combined Integral Method (CIM) (a hybrid of the TIM and RIM) is proposed. It is demonstrated that the CIM may be more accurate than a second-order perturbation up to $\beta = 7$ for a one-phase problem and $\beta = 51$ for a two-phase. It is applied to ablation in Mitchell [86].

In Hill's book [64], a wide range of integral methods, HBIM, Megerlin's method, and the catchily titled "integral formulation by integration" are tested in Cartesian, cylindrical, and spherical co-ordinates. Ribera et al. [133] apply the TIM and HBIM, in original and transformed co-ordinate systems, in both spherical and cylindrical co-ordinates. By changing the temperature variable to $u = rT$ in the spherical system, it was shown that the TIM provides highly accurate results. The choice $n = 1.6$ proved more accurate than the perturbation solution up to $\beta = 10$, when errors became negligible for both methods. Results for the cylindrical problem, via perturbation, HBIM, or TIM were less accurate, the preferred method depending on the boundary condition and Stefan number. In this case, the authors could make no definite conclusion about which method to apply.

2.3 One-Dimensional Sublimation: An Example Using a Combination of Techniques

Sublimation is the process whereby a solid is transformed immediately to a gas (without ever becoming liquid) by a high-energy source. The related term ablation is a more destructive process in that it can also involve physical destruction, for example, by erosion, or simply where the sublimated material is removed by some external force. In the case of purely thermal processes, the terms are often used interchangeably. Examples of sublimation occur with solid carbon dioxide (dry ice), which readily transforms to its gaseous state, similarly for naphthalene, which is used in mothballs. Snow and ice sublime at a slow rate. High temperatures result in sublimation or ablation in the process of laser drilling and also on heat shields on spacecraft re-entering the Earth's atmosphere.

Here we will study the process of one-dimensional sublimation in the context of a high-energy laser applied to a solid surface. Andrews and Atthey [6] first examined this problem mathematically, but it has been revisited many times. Crank

[36] discusses various limiting stages of the process, applying Laplace transforms and perturbations in his analysis. Fulford and Broadbridge [55] use it as an example in their text on case studies in diffusion. Mitchell and Myers [88] use a heat balance method.

For a semi-infinite solid, the dimensional mathematical model for ablation is given by

$$\frac{\partial T}{\partial t} = \alpha_s \frac{\partial^2 T}{\partial x^2}, \qquad \rho_s L_m \frac{ds}{dt} = k_s \frac{\partial T}{\partial x} + \frac{W}{A} \tag{2.121}$$

where W/A represents the laser energy per unit area. The boundary and initial conditions are

$$T(s,t) = T_m \quad T|_{x=\infty} = T(x,0) = T_\infty \quad s(0) = 0\,. \tag{2.122}$$

Note, in this standard analysis, it is assumed that the surface instantaneously reaches the ablation heat, while the material below is unaffected. This approximation holds as the laser energy tends to infinity. A more accurate approach involves two stages, the first where the laser energy goes into raising the surface temperature to the ablation value and the second is the actual ablation phase.

The system is nondimensionalised such that

$$\hat{T} = \frac{T - T_\infty}{T_m - T_\infty} \qquad \hat{x} = \frac{x}{L}, \qquad \hat{t} = \frac{t}{\tau}. \tag{2.123}$$

There is no natural length-scale for this problem. However, physically, it is clear that the removal of solid is driven by the laser, so we balance the first and final terms in the Stefan condition to obtain $L = W\tau/(\rho_s L_m A)$. The heat equation must then provide the time-scale $\tau = L^2/\alpha$: this is the order of magnitude of time for the heat to penetrate a depth L. Substituting for τ and rearranging the expression for L gives the length-scale $L = \rho_s L_m A\alpha/W$.

The nondimensional system (without hats) may now be written as

$$\frac{\partial T}{\partial t} = \frac{\partial^2 T}{\partial x^2} \qquad \frac{ds}{dt} = \varepsilon \left.\frac{\partial T}{\partial x}\right|_{x=s} + 1\,, \tag{2.124}$$

subject to

$$T(s,t) = 1 \quad T|_{x=\infty} = T(x,0) = 0 \quad s(0) = 0\,. \tag{2.125}$$

The small parameter $\varepsilon = k_s \Delta T/(\rho_s L_m \alpha) \sim 0.2$ for a typical laser welding process [6].

Starting with a standard perturbation, we write

$$T = T_0 + \varepsilon T_1 + \cdots \qquad s = s_0 + \varepsilon s_1 + \cdots\,. \tag{2.126}$$

To first order, the governing equations are

$$\frac{\partial T_0}{\partial t} + \varepsilon \frac{\partial T_1}{\partial t} = \frac{\partial^2 T_0}{\partial x^2} + \varepsilon \frac{\partial^2 T_1}{\partial x^2}\,, \tag{2.127}$$

$$\frac{ds_0}{dt} + \varepsilon \frac{ds_1}{dt} = \varepsilon \left.\frac{\partial T_0}{\partial x}\right|_{x=s_0} + 1\,. \tag{2.128}$$

Care must be taken with the boundary condition at the interface. To first order, this may be written as

$$\begin{aligned} T(s,t) &= T_0(s_0 + \varepsilon s_1, t) + \varepsilon T_1(s_0 + \varepsilon s_1, t) \\ &= T_0(s_0, t) + \varepsilon s_1 \left.\frac{\partial T_0}{\partial x}\right|_{x=s_0} + \varepsilon T_1(s_0, t) = 1\,. \end{aligned} \tag{2.129}$$

The initial and far-field conditions simply show that the temperature is zero at all orders.

An interesting feature of this approach is that the leading order position of the moving boundary may be calculated immediately from Eq. (2.128)

$$s_0 = t\,. \tag{2.130}$$

This is with no knowledge of the temperature field. Physically, this tells us that to leading order, the phase change is solely driven by the laser energy and the solid temperature plays a negligible role. However, given that $\varepsilon \sim 0.2$, we can expect this "negligible" role to contribute to the order 20%. This spurs us on to investigate the higher-order terms.

To calculate s_1 requires knowledge of T_0, which satisfies a standard heat equation on a moving boundary. To remedy this, we use a boundary-fixing transformation $\eta = x - s_0(t) = x - t$. Denoting $T_0(x,t) = F(\eta(x,t), t)$, the leading order heat equation is

$$\frac{\partial F}{\partial t} - \frac{\partial F}{\partial \eta} = \frac{\partial^2 F}{\partial \eta^2}\,, \tag{2.131}$$

subject to

$$F(0,t) = 1 \qquad F|_{\eta=\infty} = F(\eta, 0) = 0\,. \tag{2.132}$$

In Crank [36], this is solved exactly via Laplace transforms but here, since we are approaching the end of the approximation section, we will follow Fulford [55] and seek a large time solution. Choosing a new time-scale $t = \bar{t}/\varepsilon$, the time derivative in (2.131) becomes negligible and the equation takes on its steady-state form (to leading order in ε). The appropriate solution is

$$F = e^{-\eta}, \qquad T_0 = e^{-(x-t)}. \tag{2.133}$$

Then from the first-order terms of the Stefan condition (2.128),

$$\frac{ds_1}{dt} = \left.\frac{\partial T_0}{\partial x}\right|_{x=s_0} = \left.-e^{-(x-t)}\right|_{x=s_0} = -1. \tag{2.134}$$

Hence, $s_1 = t$ and so to first order $s = (1-\varepsilon)t$. In dimensional form, this is

$$s = \left(1 - \frac{c_s(T_m - T_\infty)}{L_m}\right)\frac{W}{\rho_s L_m A}\, t, \tag{2.135}$$

after replacing the thermal diffusivity $\alpha_s = k_s/(\rho_s c_s)$. This makes clear the fact that the sublimation process is controlled by the laser power and the density of the material. The correction term shows that the process is slowed down by heat flow in the solid, and the effect increases with decreasing solid temperature, i.e. it is harder to sublimate a cold solid than a hot one.

This same problem is tackled in Myers [100] using the TIM. The initial pre-ablation is tackled as a standard thermal problem with a constant flux at the boundary, so Eq. (2.114) defines the temperature with $n = 3.584$. This continues until $T(0,t) = 1$ (i.e. the boundary reaches the ablation temperature). Once ablation begins, the temperature is approximated by

$$T = \left(\frac{\delta - x}{\delta - s}\right)^n \qquad n > 1 \tag{2.136}$$

which satisfies $T(s,t) = 1$, $T(\delta,t) = 0$. The heat balance integral leads to a first-order ordinary differential equation for $\delta(t)$ and $s(t)$:

$$\frac{d\delta}{dt} + n\frac{ds}{dt} = \frac{n(n+1)}{\delta - s}. \tag{2.137}$$

The Stefan condition results in

$$\frac{ds}{dt} = 1 - \frac{\varepsilon n}{\delta - s}. \tag{2.138}$$

It is a simple matter to eliminate the $\delta - s$ terms resulting in an equation for δ_t and s_t, which may be immediately integrated. The resultant equation gives a relation between δ and s, which then allows one variable to be removed from the above equations, and we are left with a single first-order differential equation, which must be solved numerically. In [100], it is shown that taking $n = 3.584$ (which avoids a discontinuity between pre- and post-ablation stages) leads to results that are indistinguishable from the numerical solution of (2.124, 2.125).

2.4 A Brief Sojourn on Numerical Methods

In the following chapters, there will be no deep focus on numerical methods. This is such a large area in itself that it could easily form a separate book. Consequently, here we simply provide a very brief overview of standard techniques.

Crank [36] defines three categories of numerical techniques: front tracking, front fixing, and fixed domain. With front tracking, the heat equations are solved in each phase and the position of the boundary is determined at each time-step. In general, the front will not be located on a grid point. Various methods have been developed to deal with this, such as modifying the derivative formulae in the vicinity of the boundary or employing a variable time-step such that the boundary falls on a grid point or choosing the space interval such that $\delta x = s(t)/I$, where I is the constant number of grid points in the phase.

Front fixing methods involve choosing a new space coordinate – we have already encountered a number of such transformations in Sect. 2.2.1. The most common is to work with the variable $\eta = x/s(t)$, and then the governing equations are given by (2.52, 2.53). This removes any difficulty associated with the interface lying between grid points. However, both front tracking and front fixing methods suffer from an initialisation problem. For front tracking, one of the phases does not exist at $t = 0$ and so the associated domain has zero size. This issue translates to a singularity at $t = 0$ in the front fixing approach. One way to avoid the latter, at least when choosing $\eta = x/s(t)$, is to define the function $z = s^2$, as discussed in Sect. 2.2.1. The former has been dealt with by first seeking small time solutions and applying these as the "initial" condition for some small time $t > 0$; see Sect. 2.2.3 and [92].

With both front tracking and front fixing approaches, standard techniques, such as finite difference, finite element, boundary elements, spectral collocation, may be employed; see, for example, [23, 24, 36, 92, 95, 127, 160].

Aside from the issues mentioned above, the extension to two and three dimensions is complicated with front tracking and front fixing methods. Both require solving distinct heat equations and satisfying the Stefan condition at the boundary. Crank [36] discusses difficulties encountered when the boundary does not move smoothly or monotonically. This motivates a new formulation that applies over the whole domain and leads us to the fixed domain category. Common fixed domain approaches include the enthalpy method, effective heat capacity, level sets, and phase field methods. Having already introduced the enthalpy formulation in Sect. 1.2.2, we now focus on this.

Peters et al. [126] advocate using the enthalpy method due to the Stefan condition being treated implicitly, the governing equation being valid over the whole domain, and simple conversion to numerics. Recalling the definition of specific enthalpy

$$\mathrm{h} = \begin{cases} c_s(T - T_m) & \text{for } T < T_m \\ \in [0, L_m] & \text{for } T = T_m \\ c_l(T - T_m) + L_m & \text{for } T > T_m \,, \end{cases} \tag{2.139}$$

where we no longer distinguish between liquid and solid temperatures – the temperature is simply a variable throughout the domain, and there exists a one-to-one relation between temperature and specific enthalpy. With a constant density and thermal conductivity throughout the process, for the one-dimensional Cartesian problem, we may write

$$\rho \frac{\partial \mathrm{h}}{\partial t} = \frac{\partial}{\partial x}\left(k \frac{\partial T}{\partial x}\right) . \tag{2.140}$$

This holds throughout the domain and may be solved by standard techniques for either h or T. Fluid flow follows through the addition of the term $(\rho u \mathrm{h})_x$. The extension to two and three dimensions follows intuitively. Different thermal conductivity in either phase is dealt with through the Kirchoff transformation. For constant conductivities in each phase,

$$v(T) = \int_{T_m}^{T} k(\xi)\, d\xi = \begin{cases} k_s(T - T_m) & \text{for } T < T_m \\ 0 & \text{for } T = T_m \\ k_l(T - T_m) & \text{for } T > T_m . \end{cases} \tag{2.141}$$

There is now a one-to-one relation between T and v, and the enthalpy equation may be solved for h, T, or v. An example of the enthalpy method applied to a one-dimensional aircraft icing problem is discussed in Sect. 3.4.1. Density changes may be included through the enthalpy-porosity method, which then brings in convection; see [160]. The numerical convergence has been shown to speed up significantly in one and two dimensions through the "flag update method", where the phase of the solution is tracked through an additional variable [127]. This appears to have been introduced in [153]. Voller and Cross introduced some of the key works in the development of the enthalpy method; see [150, 151] and references therein.

Detailed discussions of other fixed domain methods may be found in [68, 130, 160].

2.5 Conclusion

Now that we are armed with a wealth of approximate and exact solution techniques, we can dive straight into more complex problems. In the next chapter, we will examine phase change where the melt region moves due to some external force. This takes us into the realm of fluid flows, a vast field in itself.

But beware, the examples provided so far have all worked out nicely. They were arranged that way. Perturbation expansions can become a whole lot more complicated when it turns out there are boundary layers or multiple balances within an equation. Similarity solutions seldom work on physically realistic problems and must therefore be combined with other, approximate techniques. Heat balance methods work well when they work, but they are not appropriate for all situations, for example, when we deal with heat flowing like a wave in the final chapter. Nondimensionalisation, when done well, will always help identify the driving forces behind a model and how to reduce it but the scaling may not be obvious or intuitive and then boundary layers may exist, requiring local rescaling. Basically, an applied mathematician must do whatever it takes to find a usable solution. It may take a lot of attempts to find the correct approach.

Chapter 2 Exercises

1. (a) In Sect. 2.1.1, the solidification of a semi-infinite liquid is analysed, leading to the classical Neumann solution. Work through the equivalent problem for the melting of a semi-infinite solid, i.e. write down the non-dimensional governing equations through to the expressions for the temperature, equivalent to (2.9) and the corresponding transcendental equation for λ, Eq. (2.12).
   Plot the solution for various values of β.
   Prove that the transcendental equation has a unique solution $\lambda > 0$ for any $\beta \geq 0$.
   (b) Repeat the process to include a density variation.
2. Reduce the system of question 1 to a one-phase problem to study the melting of a semi-infinite solid initially at the solidus.
   Is the melting faster or slower than the equivalent two-phase problem?
3. Using the Landau transformation $\eta = x/s(t)$, transform the two-phase problem with density change (the liquid heat variation is described by Eq. (2.16)) to one with a fixed boundary.
4. Consider the two-phase problem, on a finite domain $x \in [0, 1]$:

$$\frac{\partial T}{\partial t} = \frac{\partial^2 T}{\partial x^2}, \qquad \frac{\partial \theta}{\partial t} = \alpha \frac{\partial^2 \theta}{\partial x^2}, \qquad \beta \frac{ds}{dt} = \frac{\partial T}{\partial x} - k \frac{\partial \theta}{\partial x}\bigg|_{x=s}$$

   where $\theta(x, 0) = 1$, $T(0, t) = T_0 < 0$ (for $t > 0$) and $\theta_x(1, t) = 0$. For large β, rescale time and find the leading order perturbation solution for the solid in terms of s. Now try a similar approach for the liquid temperature. What goes wrong? How could you deal with this issue?
5. Ribera et al. [133] study a one-phase spherically symmetric melting problem, where a solid initially at the solidus is melted by imposing a fixed temperature at the boundary.

   (a) Working on the phase change time-scale, the system is governed by

$$\varepsilon \frac{\partial \theta}{\partial t} = \frac{1}{r^2} \frac{\partial}{\partial r}\left(r^2 \frac{\partial \theta}{\partial r}\right), \quad \theta(R, t) = 0, \quad \theta(1, t) = 1, \quad \frac{dR}{dt} = -\frac{\partial \theta}{\partial r}\bigg|_{r=R}.$$

   Calculate the leading order temperature profile in terms of $R(t)$ and consequently integrate the Stefan condition to determine an implicit solution for the evolution of the melt front.

At what time does the sphere completely melt?

(b) Now transform the system to a "Cartesian" form by setting $u = r\theta$. Write down the governing equations. Does the leading order solution coincide with the previous solution?

6. Following on from the above analysis, find the leading order solution for $R(t)$ for the equivalent cylindrical problem:

$$\varepsilon \frac{\partial \theta}{\partial t} = \frac{1}{r} \frac{\partial}{\partial r} \left(r \frac{\partial \theta}{\partial r} \right), \quad \theta(R, t) = 0, \quad \theta(1, t) = 1, \quad \frac{dR}{dt} = - \left. \frac{\partial \theta}{\partial r} \right|_{r=R}.$$

How long does it take to completely melt?

7. Consider the one-phase solidification problem defined in Sect. 2.1.3. In Fig. 2.5, the exact, perturbation, and TIM solutions are compared for $\beta = 1$. Plot your own solutions for different β cases, which is best for $\beta > 1$ and for $\beta < 1$?
8. In deriving the TIM solution of (2.1.3) before equation (2.117), the choice was made to evaluate $T_x(s, t)$ via the Stefan condition rather than the chosen expression for $T(x, t)$. Follow through the solution process, but now use the expression for T to evaluate T_x. Compare the accuracy of solutions obtained through the two approaches (this is best quantified through the value of the least-squares error).
9. Replacing the boundary condition at $x = 0$ with the Robin condition, $T_x(0, t) = T(0, t) - 1$, in the system (2.98), the TIM gives the solution:

$$T = \frac{\delta}{n + \delta} \left(1 - \frac{x}{\delta} \right)^n, \quad t = \frac{1}{n(n + 1)} \left(\frac{\delta^2}{2} + n\delta - n^2 \ln \left(\frac{n + \delta}{n} \right) \right)$$

with $n = 3.584$.

Compare this with the exact solution:

$$T = \operatorname{erfc} \left(\frac{x}{2\sqrt{t}} \right) - e^{x+t} \operatorname{erfc} \left(\frac{x + 2t}{2\sqrt{t}} \right).$$

10. Adsorption is the process where molecules attach to a surface. The process of column adsorption is used to clean contaminated fluids, for example, capturing CO_2 from air or pollutants from water. It involves passing a contaminated fluid through adsorbent particles packed into a circular cylindrical column. In this scenario, the nondimensional adsorption of trace amounts of contaminant may be described by

$$Da \frac{\partial c}{\partial t} + \frac{\partial c}{\partial x} = Pe^{-1} \frac{\partial^2 c}{\partial x^2} - \frac{\partial q}{\partial t}$$

$$\frac{\partial q}{\partial t} = 1 - q,$$

where c is the contaminant concentration and q the amount already adsorbed (or the amount of occupied adsorption sites). Here the imposed sink indicates an adsorption rate proportional to the available attachment sites. The nondimensional numbers Da, Pe^{-1} represent the Dahmköhler and inverse Peclét numbers, respectively.
The concentration and hence the amount adsorbed are zero at a moving front, $x = s(t)$, such that $c(s,t) = q(s,t) = 0$. At the inlet, contaminated material enters; this is scaled such that $c(0,t) = 1$ and, after sufficient time, the adsorbent material near the inlet will be fully occupied such that $q(0,t) = 1$:

(a) Assuming $Da, Pe^{-1} \ll 1$, use a travelling wave substitution, $\eta = x - s(t)$ such that $F(\eta) = c(x,t)$, $G(\eta) = q(x,t)$, to reduce the system to a pair of ODEs. From this system, determine the form of $s(t)$.
(b) Writing $s(t) = vt + s_0$ (which may give a hint to the previous question), solve the system subject to $F = G = 0$ at $\eta = 0$, $F, G \to 1$ as $\eta \to -\infty$ and determine the velocity v. (Note, the conditions at $\eta = 0$ also defines the position of the front as the point where all contaminant has been removed).
(c) Now repeat the process with Da included.

Finally, note we cannot determine s_0 from an initial condition – the travelling wave does not hold at $t = 0$. In practice, we take an exprimental time, measured at the column outlet to find s_0. Details are provided in [103, 120, 121].

11. The Shrinking Core Model is common to the chemical engineering literature and may be used to describe the process where solid particles are consumed through reacting with some surrounding material. It is analogous to a spherical Stefan problem, but here the reacted material remains around the core and the reactant must diffuse through it to reach the front; see [73]. The standard model is described by

$$\frac{\partial c}{\partial t} = \frac{D}{r^2}\frac{\partial}{\partial r}\left(r^2\frac{\partial c}{\partial r}\right),$$

$$c(R,t) = c_R,$$

$$D\left.\frac{\partial c}{\partial r}\right|_{r=s} = k\left(c(s,t) - c_{eq}\right),$$

$$\frac{\rho_s}{M}\frac{ds}{dt} = -\left.D\frac{\partial c}{\partial r}\right|_{r=s},$$

where the unreacted material occupies $0 \le r \le s(t)$ and $s(0) = R$ defines the original radius. The concentration of the material reacting with the solid is c, which has a constant concentration c_R at the outer boundary. This material must diffuse through the reacted section $s \le r \le R$ (and note we have assumed no density change in this region). The constants ρ_s, M, D, k represent the density of the core, the molecular mass of c, the diffusion coefficient of c through the reacted material, and a reaction constant:

(a) Define

$$\hat{r} \equiv \frac{r}{R}, \qquad \hat{s} \equiv \frac{s}{R}, \qquad \hat{t} \equiv \frac{t}{\tau}, \qquad \hat{c} \equiv \frac{c - c_{eq}}{c_R - c_{eq}},$$

(where c_{eq} is a known constant value). From this identify, the diffusion and growth time-scales. There is a third time-scale, the kinetic time-scale. Find an expression for this by replacing $D = R^2/\tau$ in the condition at $\hat{s}$.

(b) By applying a suitable assumption, determine a pseudo-steady-state solution (in the form $t = f(s)$). At what time is the core all used up?

(c) A standard reduction quoted in textbooks (e.g. see [73]) is obtained by setting $Rk/D \gg 1$, resulting in the *Diffusion Controlled* model:

$$\frac{\hat{s}^3 - 1}{3} - \frac{\hat{s}^2 - 1}{2} = D\hat{t}/(Rk)\,.$$

Is this approximation always valid? If not, when can it be applied?

Chapter 3
Solidification of a Thin Liquid Layer

In the first chapter, we briefly mentioned a model for phase change when one of the phases is moving, without elaborating on the form of the motion. In general, it can be quite tricky to determine the velocity **u**, which may be the result of external forces and internal forces or simply due to density change. In Chaps. 4 and 5, we will consider a number of special cases where the flow is due to the density change. In this chapter, we will look at the situation where the liquid flow is driven by external forces.

The flow of a viscous liquid is described by the Navier-Stokes equations. They may be used to describe complex physical phenomena such as atmospheric motion and ocean currents. They also provide many mathematical challenges concerning existence and smoothness. However, all of this is beyond the scope of the present text; our focus is on phase change. Consequently, here we will restrict the analysis to the special case of a Newtonian fluid that forms a thin layer. This permits significant simplifications to the Navier-Stokes equations and hence analytical progress.

3.1 Lubrication Theory

Equation (1.7) defines the heat equation in the presence of a mobile liquid layer. The introduction of an unknown velocity requires additional equations to quantify the fluid flow. To achieve this, we begin by writing a standard form of the Navier-Stokes equation:

$$\rho_l \left(\frac{\partial \mathbf{u}}{\partial t} + \mathbf{u} \cdot \nabla \mathbf{u} \right) = -\nabla p + \nabla \cdot (\mu_l \nabla \mathbf{u}) \ , \tag{3.1}$$

where p is the fluid pressure and μ_l the dynamic viscosity. In this chapter, we will focus solely on Newtonian fluids and so take μ_l to be constant. If the fluid

T. G. Myers, *Theoretical and Practical Stefan Problems*, Interdisciplinary Applied Mathematics 63, https://doi.org/10.1007/978-3-032-04826-4_3

is incompressible, then

$$\nabla \cdot \mathbf{u} = 0 . \tag{3.2}$$

Assuming laminar flow in the $x - z$ plane, we nondimensionalise the fluid equations such that

$$\hat{x} = \frac{x}{L}, \quad \hat{z} = \frac{z}{H}, \quad \hat{u} = \frac{u}{U}, \quad \hat{w} = \frac{w}{W}, \quad \hat{p} = \frac{p - p_a}{P}, \quad \hat{t} = \frac{t}{\tau_f}, \tag{3.3}$$

where L, H are typical length-scales in the x, z directions, respectively. The corresponding velocity scales are U, W. The pressure scale P depends on the physical problem and p_a represents the ambient pressure. The flow time-scale $\tau_f = L/U$ is simply the time taken for the fluid to flow the distance of the length-scale. The essence of lubrication theory is that the length-scale perpendicular to the surface is much smaller than that along the surface $H \ll L$; this defines the key small parameter $\varepsilon = H/L \ll 1$. To balance terms in the incompressibility equation shows that $W = HU/L$, that is, the flow in the z direction is much slower than in the x direction. Lubrication theory permits flow in both the x and z directions; a more restrictive approximation is to define $w = 0$ everywhere, which is termed unidirectional flow. In the case where a thin layer of fluid flows over a freezing substrate, the density change upon solidification will force a vertical velocity; for this reason, we prefer lubrication rather than unidirectional flow (although the differences will be small).

The choice $W = \varepsilon U$ results in identical dimensional and nondimensional versions of the incompressibility condition:

$$\frac{\partial \hat{u}}{\partial \hat{x}} + \frac{\partial \hat{w}}{\partial \hat{z}} = 0 . \tag{3.4}$$

The scaled x component of (3.1) is

$$\frac{\rho_l U}{\tau_f} \frac{\partial \hat{u}}{\partial \hat{t}} + \frac{\rho_l U^2}{L} \left(\hat{u} \frac{\partial \hat{u}}{\partial \hat{x}} + \hat{w} \frac{\partial \hat{u}}{\partial \hat{z}} \right) = - \frac{P}{L} \frac{\partial \hat{p}}{\partial \hat{x}} + \frac{\mu_l U}{H^2} \left(\frac{H^2}{L^2} \frac{\partial^2 \hat{u}}{\partial \hat{x}^2} + \frac{\partial^2 \hat{u}}{\partial \hat{z}^2} \right) . \tag{3.5}$$

Setting $\tau_f = L/U$, an appropriate rearrangement gives

$$\varepsilon^2 \frac{\rho_l U L}{\mu_l} \left(\frac{\partial \hat{u}}{\partial \hat{t}} + \hat{u} \frac{\partial \hat{u}}{\partial \hat{x}} + \hat{w} \frac{\partial \hat{u}}{\partial \hat{z}} \right) = - \varepsilon^2 \frac{PL}{\mu_l U} \frac{\partial \hat{p}}{\partial \hat{x}} + \varepsilon^2 \frac{\partial^2 \hat{u}}{\partial \hat{x}^2} + \frac{\partial^2 \hat{u}}{\partial \hat{z}^2} . \tag{3.6}$$

Lubrication theory is based on the requirement that both the square of the aspect ratio and the reduced Reynolds number (where the Reynolds number $Re = \rho_l U L/\mu_l$) are small ε^2, $\varepsilon^2 Re \ll 1$, and hence most of the terms in the above equations are negligible. When dealing with the added complication of solidification, the interesting behaviour occurs over the time-scale of the phase change, which is invariably longer than the flow time-scale. This only affects the time derivative term in (3.5), which plays an even smaller role, and so the left-hand side remains negligible on this time-scale as well. The pressure term suggests the scale $P = \mu_l U/(\varepsilon^2 L)$. The presence of the ε^2 term in the denominator indicates that lubrication pressures are very high and explains why thin fluid films can support large weights (think how a lubricated vehicle axle supports the vehicle's weight). Dropping the hat notation and applying the same scaling to the w component, we obtain the reduced form of the flow equations in the x and z directions:

$$\frac{\partial p}{\partial x} = \frac{\partial^2 u}{\partial z^2} + \mathscr{O}(\varepsilon^2, \varepsilon^2 Re)\,, \qquad \frac{\partial p}{\partial z} = \mathscr{O}(\varepsilon^2, \varepsilon^2 Re)\,. \tag{3.7}$$

Neglecting small terms, the second equation shows that the pressure is independent of z, and consequently the first equation can be integrated immediately to determine the velocity u:

$$u = A_2 \frac{\partial p}{\partial x}\frac{z^2}{2} + A_1 z + A_0\,, \tag{3.8}$$

where the constants A_n are determined from the boundary conditions. The velocity w then follows from integrating the incompressibility condition (3.4) with respect to z.

Where a liquid layer contacts a solid, the most commonly applied boundary condition is that of no-slip, which means that the fluid velocity matches that of the solid. If a stationary substrate is aligned parallel to the x-axis, then no-slip indicates:

$$\mathbf{u}\cdot\hat{\mathbf{t}} = \mathbf{u}\cdot\hat{\mathbf{n}} = 0\,, \quad \Rightarrow \quad u = w = 0\,, \tag{3.9}$$

where $\mathbf{u} = U(u, \varepsilon w)$, $\hat{\mathbf{n}} = (0, 1)$, $\hat{\mathbf{t}} = (1, 0)$ are the unit normal and unit tangent vectors to the x-axis. If the substrate has a varying height, for example, at a space-dependent phase change boundary, defined by $z = s(x, t)$, then the requirement of zero tangential velocity becomes

$$\mathbf{u}\cdot\hat{\mathbf{t}} = 0\,, \quad \text{where} \quad \hat{\mathbf{t}} = \frac{(1, \varepsilon s_x)}{\sqrt{1+\varepsilon^2 s_x^2}} \approx (1, \varepsilon s_x) + \mathscr{O}(\varepsilon^2)\,, \tag{3.10}$$

and therefore

$$\mathbf{u}\cdot\hat{\mathbf{t}} = \frac{U}{\sqrt{1+\varepsilon^2 s_x^2}}(u + \varepsilon^2 s_x w) = 0 \quad \Rightarrow \quad u + \mathscr{O}(\varepsilon^2) = 0\,, \tag{3.11}$$

at $z = s(x, t)$. So, within the restrictions of lubrication theory, zero tangential velocity at the interface still requires $u(x, s, t) = 0$.

One popular modification of the no-slip condition at $z = 0$ is the slip condition:

$$u = L_s \frac{\partial u}{\partial z} \tag{3.12}$$

where L_s is the slip length. This condition indicates that the velocity $u(x, 0, t) > 0$ and only reaches zero at a distance L_s below the substrate. The concept of slip is an old one, attributed to Navier [154], and has many practical uses. When dealing with flow over a rough surface, it permits a boundary condition to be applied on a flat surface, and then the slip length is related to the roughness height. In thin-film theory, it removes the stress singularity at the moving contact line, where the fluid front moves over a dry surface. Myers [101] uses experimental observations of a *depletion layer* (a region where, due to the low concentration of molecules, the viscosity is very low) near the wall of a carbon nanotube to motivate a slip length and hence explain the enhanced flow of water in carbon nanotubes. We will return to the concept of slip in Chap. 6.

At a free fluid surface, $z = h(x, t)$, continuity of stress is applied. Subject to the restrictions of lubrication theory, the fluid stress tensor is written in numerous textbooks; see, for example, [2, 122]. To understand the normal stress component requires knowledge of the Young-Laplace equation. This relates the pressure jump across the interface between two static fluids to the surface tension, $\Delta p = -\sigma\kappa$, where κ is the curvature and σ the surface tension. Within the restrictions of lubrication theory, where gradients are small, the curvature is simply the Laplacian of the free surface shape (to $\mathcal{O}(\varepsilon^2)$). After neglecting terms of $\mathcal{O}(\varepsilon^2)$, continuity of shear and normal stress at a liquid-gas interface then reduces to

$$\frac{\partial u}{\partial z} = \frac{\mu_g U_g}{\mu_l U}\frac{\partial u_g}{\partial z}, \qquad p = -C\frac{\partial^2 h}{\partial x^2}, \tag{3.13}$$

where μ_g, U_g are the viscosity and velocity of the overlying fluid, the inverse capillary number $C = \varepsilon^3\sigma/\mu_l U$, and σ is the surface tension between the two fluids. If the fluid above is a gas, where $\mu_g \ll \mu_l$, then the stress imposed by the gas is often neglected (unless its velocity is high). The second condition indicates that the pressure jump at a free surface is caused by the surface tension, and this stress is proportional to the curvature. A further condition at a free surface is that fluid particles generally remain on the surface (unless, e.g. when there is phase change, such as evaporation), resulting in the *kinematic condition*. Mathematically, this may be stated as

$$\frac{d}{dt}(z - h(x, t)) = 0, \quad \Rightarrow \quad \frac{\partial}{\partial t}(z - h) + \mathbf{u}\cdot\nabla(z - h) = 0. \tag{3.14}$$

Substituting for the components of $\mathbf{u}$ leads to

$$w = \frac{\partial h}{\partial t} + u\frac{\partial h}{\partial x}, \tag{3.15}$$

which is applied at the free surface $z = h(x, t)$. An alternative approach to derive this equation is to apply a mass balance. This approach is particularly useful when a phase change occurs at the interface. To illustrate this, consider an air-water interface, where the air contains water vapour. The dimensional mass balance requires

$$\rho_l \hat{\mathbf{n}} \cdot (\mathbf{u} - \mathbf{u}_b) = \rho_a \hat{\mathbf{n}} \cdot \left(\mathbf{u}_g - \mathbf{u}_b\right) \tag{3.16}$$

where $\mathbf{u} = U(u, \varepsilon w)$, $\hat{\mathbf{n}} = (-\varepsilon h_x, 1)/\sqrt{1+\varepsilon^2 h_x^2} \approx (-\varepsilon h_x, 1)$, while the boundary velocity is $\mathbf{u}_b = (0, Hh_t/\tau_f)$. Note, the density ρ_a is that of the material available for transfer in the overlying gas phase, which is not the same as the density of the gas phase itself! For example, in the aircraft icing study described in Sect. 3.4 where growth is due to incoming water droplets, $\rho_a = \phi \rho_w$ where $\phi \ll 1$ is the volume fraction of water droplets in the air and ρ_w the water density. Equation (3.16) leads to

$$w = \frac{L}{U\tau_f}\left(1 - \frac{\rho_a}{\rho_l}\right)\frac{\partial h}{\partial t} + u\frac{\partial h}{\partial x} + \frac{\rho_a U_g}{\varepsilon \rho_l U}\hat{\mathbf{n}} \cdot \mathbf{u}_g\,. \tag{3.17}$$

Firstly, we note that $\tau_f = L/U$ was defined earlier so that on the flow time-scale $L/(U\tau_f) = 1$. We leave it as τ_f in the equations in case we wish to change time-scale later. Although, in general, $\rho_a/\rho_l \ll 1$ may be neglected in the first term on the right-hand side, the coefficient of the final term has ε in the denominator. Hence, despite the small value of the density ratio, for gas velocities of the order $U_g = \varepsilon \rho_l U/\rho_a$, this term can affect the mass balance, leading to a growing fluid layer. In the limit $\rho_a \to 0$, the standard kinematic condition (3.15) is retrieved. This limit makes it clear that the statement that fluid particles remain on the surface is equivalent to zero transfer to the overlying fluid.

A similar analysis at a solid substrate $z = s(x, t)$ gives

$$\rho_l \hat{\mathbf{n}} \cdot (\mathbf{u} - \mathbf{u}_b) = \rho_s \hat{\mathbf{n}} \cdot (\mathbf{u}_s - \mathbf{u}_b)\ , \tag{3.18}$$

where $\hat{\mathbf{n}} \approx (-\varepsilon s_x, 1)$, $\mathbf{u} = U(0, \varepsilon w)$ is the fluid velocity at the boundary (recall $u(x, s, t) = 0$), $\mathbf{u}_b = (0, Hs_t/(U\tau_f))$ is the boundary velocity, and $\mathbf{u}_s = \mathbf{0}$ is the solid velocity. This leads to an expression for the vertical velocity at the moving boundary:

$$w(x, s, t) = \left(1 - \frac{\rho_s}{\rho_l}\right)\frac{L}{U\tau_f}\frac{\partial s}{\partial t}\,, \tag{3.19}$$

where terms of $\mathcal{O}(\varepsilon^2)$ have been neglected. If the substrate is fixed, i.e. with no phase change, then $s_t = 0$ and so $w = 0$, as in Eq. (3.9). If the fluid and solid densities are the same, then there is no expansion or contraction upon phase change and the vertical fluid velocity is zero. Otherwise, $w \neq 0$ and the fluid must move to accommodate the new material, its velocity determined by the density difference and solidification rate.

3.1.1 Channel Flow

Before leaping in to modelling the phase change of a flowing liquid, we first analyse two common types of flow that may be described by the lubrication approximation. Later we will adapt them to include solidification.

Poiseuille or Hagen-Poiseuille flow describes the pressure-driven laminar flow through a channel or pipe. With straight, parallel boundaries, the equations may be derived from Navier-Stokes by assuming a velocity profile $\mathbf{u} = (u, 0)$ (which is the unidirectional flow mentioned in the previous section). If the channel is long and thin, the same results may be achieved, within $\mathcal{O}(\varepsilon^2, \varepsilon^2 Re)$, using lubrication theory. However, the advantage of lubrication theory is that if the sides are not exactly parallel, such that unidirectional flow no longer holds, lubrication theory may still be applied (provided the channel slope is $\mathcal{O}(\varepsilon)$). To demonstrate this, we now define a Cartesian channel such that the walls are located at $z = \pm s(x, t)$. Integrating the first of Eqs. (3.7), subject to no-slip boundary conditions, $u(x, \pm s, t) = 0$, leads to

$$u = \frac{p_x}{2}(z^2 - s^2)\,. \tag{3.20}$$

The pressure gradient may be related to the fluid flux via

$$Q = \int_{-s}^{s} u\,dz = -\frac{2}{3}s^3 p_x\,. \tag{3.21}$$

So, provided we know the amount of fluid passing through the channel, we can determine the pressure gradient. Conversely, a known pressure gradient determines the flux.

Integrating the incompressibility condition, (3.4) gives an expression for the vertical velocity:

$$w(x, z, t) - w(x, 0, t) = -\int_0^z \frac{\partial u}{\partial x}\,dz\,, \tag{3.22}$$

after imposing symmetry such that $w|_{z=0} = 0$. Applying Leibniz' theorem, Eq. (2.102), (but now with an x derivative) with $a(x) = 0$, $b(x) = z$, and $f(x, z) = u(x, z)$, we may take the derivative outside of the integral in (3.22) and so

$$w(x, z, t) = -\frac{\partial}{\partial x}\int_0^z u\,dz = -\frac{\partial}{\partial x}\left[\frac{p_x}{6}(z^3 - 3s^2 z)\right]. \tag{3.23}$$

The position s is still unknown. If we equate the value of $w(x, s, t)$ obtained from Eq. (3.23) with that of Eq. (3.19), we obtain

$$\left(1 - \frac{\rho_s}{\rho_l}\right) \frac{L}{U\tau_f} \frac{\partial s}{\partial t} = \frac{\partial}{\partial x}\left[\frac{p_x}{3} s^3\right] = -\frac{1}{2}\frac{\partial Q}{\partial x}. \tag{3.24}$$

This provides an equation for s_t in terms of the pressure gradient or flux. To take the analysis further, it is best to provide a concrete example, so in Sect. 3.2, we will expand these results to deal with solidification in a microchannel.

3.1.2 Free Surface Flow

In Sect. 3.4, we consider the growth of ice on a solid substrate due to a thin layer of flowing water, where the flow is driven by air shear, gravity, and surface tension. Since the upper surface of the fluid is not constrained, this is a form of free surface flow. Free surface flow is a classical problem in fluid dynamics and, in this case, there is the added complication of phase change. Defining the position of the substrate as $z = 0$, the ice surface as $z = s(x, y, t)$, and the liquid film surface as $z = s + h$, we may apply a no-slip boundary condition at the ice surface, while at the free surface, the shear stress is due to the overlying air flow:

$$u(x, s, t) = 0, \qquad \left.\frac{\partial u}{\partial z}\right|_{z=s+h} = A. \tag{3.25}$$

The second condition derives from the first of Eqs. (3.13), where the value of the gas shear stress, A, comes from some external calculation for the gas flow.

Gravity is a body force and so augments the governing lubrication equations (3.7). The simplest way to achieve this in the above system is by simply replacing the pressure gradient with pressure gradient plus gravity $p_x \to p_x - B\hat{\mathbf{g}}\cdot\hat{\mathbf{x}}$, $p_z \to p_z - B\hat{\mathbf{g}}\cdot\hat{\mathbf{z}}$ where $B = \varepsilon^2 \rho_l g L^2/\mu_l U$ is the ratio of gravitational to viscous forces. The second of Eqs. (3.7) may now be integrated:

$$p = B\hat{\mathbf{g}}\cdot\hat{\mathbf{z}}\, z + f(x, t). \tag{3.26}$$

Applying the free surface condition (3.13) (but at $z = h + s$), which relates pressure to surface tension, determines the unknown function:

$$p = B\hat{\mathbf{g}}\cdot\hat{\mathbf{z}}(z - (h + s)) - C\frac{\partial^2}{\partial x^2}(s + h). \tag{3.27}$$

Integrating the first of Eqs. (3.7), with gravity included and subject to the boundary conditions (3.25), leads to

$$u = \frac{1}{2}\left(p_x - B\hat{\mathbf{g}} \cdot \hat{\mathbf{x}}\right)\left(z^2 - s^2 - 2(s+h)(z-s)\right) + A(z-s)\,. \tag{3.28}$$

In free surface flows, a key issue is, unsurprisingly, the position of the free surface. This may be calculated using the mass balances discussed earlier. To apply these requires knowledge of the velocities. At the ice-water interface, the vertical velocity is given by (3.19). At the liquid-air interface, the vertical velocity is specified by a modified version of Eq. (3.17), which is derived using the new interface position $z = s + h$ and corresponding normal and boundary velocity:

$$\mathbf{u}_{b2} = (0, (s+h)_t)\,, \qquad \mathbf{n}_{b2} = (-(s+h)_x, 1)\,. \tag{3.29}$$

This leads to a modified version of the kinematic condition (3.17):

$$w(x, s+h, t) = \left(1 - \frac{\rho_a}{\rho_l}\right)\frac{L}{U\tau_f}(s+h)_t + u(s+h)_x + \frac{\rho_a U_g}{\varepsilon\rho_l U}\mathbf{u}_g \cdot \mathbf{n}_{b2}\,. \tag{3.30}$$

When integrating the continuity equation across the liquid film Leibniz theorem, Eq. (2.102), allows us to write

$$w(x, z, t) - w(x, s, t) = -\int_s^z \frac{\partial u}{\partial x}\,dz = -\frac{\partial}{\partial x}\int_s^z u\,dz\,. \tag{3.31}$$

The final equality holds since no-slip dictates $u(x, s, t) = 0$. Evaluating the integral after substituting for u via (3.28) and the boundary velocity $w(x, s, t)$ via Eq. (3.30) leads to the mass balance:

$$\frac{L}{U\tau_f}\frac{\partial h}{\partial t} + \frac{\partial Q}{\partial x} = \frac{\dot{m}}{\rho_l} - \frac{\rho_s}{\rho_l}\frac{L}{U\tau_f}\frac{\partial s}{\partial t}\,, \tag{3.32}$$

where $\dot{m} = \rho_a U_g \mathbf{u}_g \cdot \mathbf{n}_{b2}/\varepsilon U$ represents the mass transfer rate at the free surface and the fluid flux:

$$Q = -\frac{h^3}{3}\left(\frac{\partial p}{\partial x} - B\hat{\mathbf{g}} \cdot \hat{\mathbf{x}}\right) + \frac{h^2}{2}A\,. \tag{3.33}$$

Study this equation well; it is key to subsequent models in this chapter and also for the modelling of solidification with a flowing material in general. It demonstrates that the fluid film height changes due to flow, represented by the change in flux, the material coming in from (or moving out to, depending on the sign of $\dot{m}$) the overlying layer, and also the mass being lost at the lower surface through conversion to a solid. Choosing the flow time-scale $\tau_f = L/U$ and setting the right-hand side of Eq. (3.32) to zero (i.e. neglecting any mass loss or gain terms), this reduces to a standard free surface thin-film equation. Since the pressure involves a second-order term and the flow equation contains $Q_x \sim p_{xx}$, this is a fourth-order nonlinear degenerate partial differential equation. It has a wide variety of physical applications (e.g. with coatings, condensate motion, heat exchangers, droplet spreading, and the ever-fascinating paint drying) and rich mathematical behaviour. The flux expression shows that the fluid motion is due to pressure gradients, gravity, and air shear. A review of similar equations, their practical uses, and mathematical properties may be found in Craster and Matar [37], Myers [96], and Oron et al. [124].

3.2 Solidification in a Microchannel

Microfluidic valves are often miniature versions of their larger counterparts. However, due to obvious constraints, such as manufacturing difficulties on a small-scale or lack of space in the microfluidic device, there are also a number of valves that exploit technology not appropriate on a larger scale. One such device is the phase change valve. The phase change valve works by freezing the fluid within the channel. They have the advantage of containing no moving parts and occupy no space within the channel but suffer from the drawback of slow response times, on the order of seconds, as opposed to milliseconds for standard valves. However, they do provide a versatile alternative to conventional microvalves when rapid closure is not critical. Practically, phase change valves have been employed in electrophoresis and liquid chromatography processes as well as micro-PCR chips. They have been studied in the field of cryopreservation, while lava solidification provides a mathematically similar (but somewhat more physically daunting) process; see [75, 109, 110, 152].

Consider a fluid flowing through a two-dimensional channel (or at least so wide in the y direction that edge effects are negligible) where the x axis follows the centre of the channel and $z \in [-H, H]$ is perpendicular to this. The configuration is shown in Fig. 3.1. We impose an initially constant fluid temperature, constant phase change, and wall temperatures and symmetry along the centreline:

$$\theta(x, z, 0) = T_0 \,, \qquad \theta(x, s, t) = T_m \,, \qquad \left.\frac{\partial \theta}{\partial z}\right|_{z=0} = 0 \,, \tag{3.34}$$

$$T(x, s, t) = T_m \,, \qquad T(x, H, t) = T_w \,, \tag{3.35}$$

where the wall temperature $T_w < T_m < T_0$. Fluid entering at the channel inlet matches the initial fluid temperature, such that $\theta(0, z, t) = T_0$. The flow is subject to no-slip and no-penetration at the liquid-solid interface as well as symmetry along the centreline. Since freezing starts at the wall, the position of the interface $s(x, 0) = \pm H$. At the inlet and outlet, the pressure is p_{in}, p_{out} with $p_{in} > p_{out}$. We

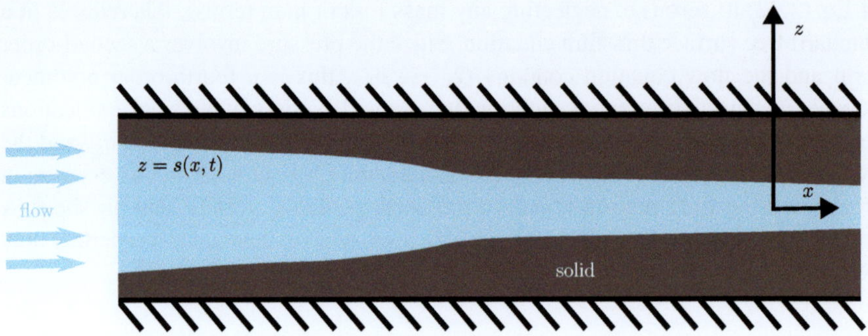

Fig. 3.1 Configuration for solidification in a channel. Fluid flows in from the left and solidifies due to the effect of the walls, located at $z = \pm H$, which are maintained below the phase change temperature. The solidification front is defined by $z = s(x, t)$

nondimensionalise according to

$$\hat{T} = \frac{T - T_m}{T_m - T_w}, \qquad \hat{\theta} = \frac{\theta - T_m}{T_0 - T_m}, \qquad \hat{x} = \frac{x}{L}, \qquad \hat{z} = \frac{z}{H}, \tag{3.36}$$

$$\hat{s} = \frac{s}{H}, \qquad \hat{t} = \frac{t}{\tau}, \qquad \hat{p} = \frac{p - p_{out}}{p_{in} - p_{out}}, \tag{3.37}$$

where τ is the solidification time-scale and immediately drop the hats.

Since the growth is two dimensional, we take the form of Stefan condition given by Eq. (1.6c). In nondimensional form, this may be written as

$$\frac{\rho_s L_m H}{\tau} \frac{\partial s}{\partial t} = \left[\frac{k_s \Delta T}{H} \left(\varepsilon \frac{\partial T}{\partial x}, \frac{\partial T}{\partial z} \right) - \frac{k_l \Delta \theta}{H} \left(\varepsilon \frac{\partial \theta}{\partial x}, \frac{\partial \theta}{\partial z} \right) \right] \cdot \hat{\mathbf{n}}, \tag{3.38}$$

(evaluated at the interface) where the unit normal

$$\hat{\mathbf{n}} = \frac{(-\varepsilon s_x, 1)}{\sqrt{1 + \varepsilon^2 s_x^2}} \approx (-\varepsilon s_x, 1) + \mathcal{O}(\varepsilon^2). \tag{3.39}$$

Choosing the solidification time-scale $\tau = \rho_s L_m H^2/(k_s \Delta T)$ (based on heat flow through the solid driving the process), where $\Delta T = T_m - T_w$ gives

$$\frac{\partial s}{\partial t} = \frac{\partial T}{\partial z} - K \frac{\partial \theta}{\partial z} + \mathcal{O}(\varepsilon^2), \tag{3.40}$$

(evaluated at the interface) where $K = k_l \Delta\theta/(k_s \Delta T)$. The heat equations (1.6a, 1.7) may be written as

$$\frac{1}{\beta}\frac{\partial T}{\partial t} = \frac{\partial^2 T}{\partial z^2} + \mathcal{O}(\varepsilon^2), \qquad s(x,t) < |z| < 1 \tag{3.41a}$$

$$\frac{1}{\beta}\frac{\partial \theta}{\partial t} + \mathrm{Pe}'\left(u\frac{\partial \theta}{\partial x} + w\frac{\partial \theta}{\partial z}\right) = \alpha\frac{\partial^2 \theta}{\partial z^2} + \mathcal{O}(\varepsilon^2), \qquad -s(x,t) < z < s(x,t). \tag{3.41b}$$

where $\alpha = \alpha_l \Delta\theta/(\alpha_s \Delta T)$, $\beta = L_m/(c_s \Delta T)$, while the reduced Peclet number $\mathrm{Pe}' = UH^2/(\alpha_s L)$ shows the relative importance of heat flow by advection or conduction to diffusion. The velocity components u, w in Eq. (3.41b) are given by Eqs. (3.20, 3.23).

Despite the simplifications, in accordance with lubrication theory, the set of governing equations (3.24, 3.41) is still rather complicated, and to make matters worse, they hold over an unknown domain, which is specified by Eq. (3.40). A further reduction may be achieved by noting that the flow time-scale is typically much smaller than the freezing time-scale, $\tau_f \gg L/U$, except for in cases of very slow flow or very fast freezing (for this reason, Eqs. (3.7) are still valid). The term involving $L/(u\tau_f) \ll 1$ in Eq. (3.24) is therefore negligible, and so the left-hand side may be dropped (physically the growth rate $s_t \ll Q_x$, while solid growth is retained in (3.40) where the terms balance). The right-hand side then indicates that the flux $Q = Q(t)$ is a function solely of time (to $\mathcal{O}(L/(u\tau_f))$. A further, and rather obvious, simplification is that the Stefan number is large, thus leading to pseudo-steady forms for the heat equations. Finally, we may reduce the equations using the Landau transformation $\bar{z}(x,z,t) = z/s(x,t)$; see [36, p187].

Applying the Landau transformation to Eq. (3.20) and replacing the pressure gradient with the flux, via Eq. (3.21), leads to

$$u = \frac{p_x s^2}{2}(\bar{z}^2 - 1) = -\frac{3Q}{4s}(\bar{z}^2 - 1). \tag{3.42}$$

Similarly for the velocity in the z direction, using Eqs. (3.21, 3.23),

$$w(x,\bar{z},t) = -\frac{\partial}{\partial x}\left[\frac{p_x s^3}{6}\left(\bar{z}^3 - 3\bar{z}\right)\right] = \frac{Q}{4}\frac{\partial}{\partial x}\left[\left(\bar{z}^3 - 3\bar{z}\right)\right]. \tag{3.43}$$

Noting that $\bar{z}_x = -\bar{z}s_x/s$, we may relate the two velocity components:

$$w(x,\bar{z},t) = -\frac{3Q}{4s}\bar{z}s_x\left(\bar{z}^2 - 1\right) = \bar{z}s_x u. \tag{3.44}$$

The new functional dependence of the variables affects the derivatives:

$$\theta(x,z,t) \to \theta(x,\bar{z}(x,z,t),t) \tag{3.45}$$

$$\Rightarrow \frac{\partial\theta}{\partial x} \to \frac{\partial\theta}{\partial x} + \frac{\partial\theta}{\partial\bar{z}}\frac{\partial\bar{z}}{\partial x}, \qquad \frac{\partial\theta}{\partial z} \to \frac{\partial\theta}{\partial\bar{z}}\frac{\partial\bar{z}}{\partial z}, \tag{3.46}$$

where $\bar{z}_x = -zs_x/s^2 = -\bar{z}s_x/s$, $\bar{z}_z = 1/s$. Neglecting terms of $\mathcal{O}(1/\beta)$, the heat flow in the liquid, described by Eq. (3.41b), becomes

$$\mathrm{Pe}'\left[u\left(\frac{\partial\theta}{\partial x} - \frac{\bar{z}s_x}{s}\frac{\partial\theta}{\partial\bar{z}}\right) + \frac{w}{s}\frac{\partial\theta}{\partial\bar{z}}\right] = \frac{1}{s^2}\frac{\partial^2\theta}{\partial\bar{z}^2}. \tag{3.47}$$

Replacing w via the relation (3.44), the two $\theta_{\bar{z}}$ terms cancel out. Then substituting for u via Eq. (3.42) and rearranging, we obtain

$$-\frac{3Q\mathrm{Pe}'}{4}\frac{\partial\theta}{\partial x}s = \frac{1}{(\bar{z}^2-1)}\frac{\partial^2\theta}{\partial\bar{z}^2}. \tag{3.48}$$

This shows the beauty of the Landau transformation; by defining a co-ordinate such that the solidification boundary is $\bar{z} = \pm 1$, there is no "vertical" fluid flow and the w term drops out. However, it does not completely eliminate the vertical heat transfer since diffusion remains.

The thermal boundary conditions are

$$T = -1 \quad \text{at} \quad \bar{z} = \frac{1}{s}, \qquad \theta = T = 0 \qquad \text{at} \quad \bar{z} = 1, \tag{3.49}$$

$$\theta_z = 0 \quad \text{at} \quad \bar{z} = 0, \qquad \theta = 1 \qquad \text{at} \quad x = 0. \tag{3.50}$$

The pseudo-steady approximation for the solid temperature leads to a trivial solution (after applying the conditions at $\bar{z} = 1, 1/s$):

$$\frac{\partial^2 T}{\partial\bar{z}^2} = 0, \quad \Rightarrow \quad T = -\frac{s(\bar{z}-1)}{1-s}. \tag{3.51}$$

Noting that the boundary position $s = s(x,t)$, Eq. (3.48) may be solved by using separation of variables. Setting $\theta = X(x)Z(\bar{z})$, we find

$$-\frac{s(x,t)}{\gamma(t)}\frac{X'}{X} = \frac{1}{(\bar{z}^2-1)}\frac{Z''}{Z} = \lambda^2, \tag{3.52}$$

where $\gamma(t) = 4/(3Q(t)\mathrm{Pe}')$. This leads to

$$\theta = \sum_{n=1}^{\infty} A_n e^{-\lambda_n^2\gamma F(x,t)} Z_n(\bar{z}), \tag{3.53}$$

where $F(x,t) = \int_0^x 1/s(\xi,t)\,d\xi$. The eigenfunctions are

$$Z_n = \frac{W_m(\lambda_n/4, 1/4, \lambda_n \bar{z}^2)}{\sqrt{\bar{z}}} + C\frac{W_w(\lambda_n/4, 1/4, \lambda_n \bar{z}^2)}{\sqrt{\bar{z}}}, \tag{3.54}$$

where the constant

$$C = \frac{\Gamma((1-\lambda_n)/4)}{2\sqrt{\pi}} = -\left(\frac{3+\lambda_n}{8\sqrt{\pi}}\right)\Gamma\left(-\frac{3+\lambda_n}{4}\right), \tag{3.55}$$

Γ is the gamma function and W_m, W_w are the Whittaker M and W functions, respectively (see, e.g. [1]). The coefficients A_n are obtained through orthogonality:

$$A_n = \frac{\int_0^1 (1-\bar{z}^2) Z_n \, d\bar{z}}{\int_0^1 (1-\bar{z}^2) Z_n^2 \, d\bar{z}}, \tag{3.56}$$

while the eigenvalues, λ_n, satisfy the transcendental equation:

$$W_m(\lambda_n/4, 1/4, \lambda_n) + C W_w(\lambda_n/4, 1/4, \lambda_n) = 0\,. \tag{3.57}$$

Both the solid and fluid temperatures are now specified in terms of the unknown s. To determine this, we apply the Stefan condition (3.40):

$$s\frac{\partial s}{\partial t} = \left(\frac{\partial T}{\partial \bar{z}} - K\frac{\partial \theta}{\partial \bar{z}}\right)\bigg|_{\bar{z}=1}\,. \tag{3.58}$$

From this, we obtain a single first-order integro-differential equation for the position of the freezing front: $s(x, t)$

$$s\frac{\partial s}{\partial t} = -\frac{s}{1-s} - K\sum_{n=1}^{\infty} A_n e^{-\lambda_n^2 \gamma(t) F(x,t)} \left.\frac{\partial Z_n}{\partial \bar{z}}\right|_{\bar{z}=1} \tag{3.59}$$

with initial condition $s(x, 0) = 1$. The value of γ appearing in the exponential depends on the flux, $\gamma \propto 1/Q$. Noting that $p_x = -3Q/(2s^3)$ and assuming a flow driven by an inlet and outlet pressure, $p|_{x=0} = 1$, $p|_{x=1} = 0$, we may integrate to determine

$$p|_{x=1} - p|_{x=0} = -1 = -\frac{3}{2}Q\int_0^1 \frac{1}{s(x,t)^3}\,dx\,, \tag{3.60}$$

which holds until the pipe closes. This provides the expression for $Q(t)$ required in Eq. (3.59).

Equation (3.59) must be solved numerically. However, this is a significantly simpler task than dealing with a system of two heat equations and two flow equations that all apply over an unknown moving domain. Once s is determined numerically, the pressure, temperature, and velocity fields may be determined from the appropriate equations. A numerical scheme is presented in Myers and Low [109, 110].

Examining the behaviour of Eq. (3.59) in an informal way, we note that the initial condition $s = 1$ indicates that the first term on the right-hand side of (3.59) drives the solidification process at early times; hence,

$$s \frac{\partial s}{\partial t} \approx -\frac{s}{1-s} . \tag{3.61}$$

Imposing the initial condition determines $s \approx 1 - \sqrt{2t}$. This has an initial infinite velocity $s_t = -1/\sqrt{2t}$, which is to be expected with the fixed temperature boundary condition (as discussed in previous chapters). So, at early times, the solidification front is independent of x (except for at the inlet where we expect a boundary layer). In fact, this approximation may hold for larger times, whenever the summation term is negligible, for example, if $K = k_l \Delta\theta/(k_s \Delta T) \ll 1$ or $\gamma F \gg 1$. The condition on K applies if the incoming fluid temperature is close to the phase change temperature or the wall temperature is much lower than the phase change temperature. For the second condition, γ large implies a small reduced Péclet number or small flux, both of which are attained with a narrow channel or slow-moving fluid: F large requires either s small or that we are sufficiently far down the channel that the water temperature is close to freezing.

In Fig. 3.2, the (nondimensional) evolution of an ice layer in a microchannel is shown. This follows from the numerical solution of the integro-differential equation (3.59) and reproduces the results presented originally in [109]. The solutions correspond to an incoming water temperature of 283 K, a fixed boundary temperature 263 K, a freezing temperature of 273.15 K, and a range of times (increasing time has s decreasing). The resultant temperature scale $\Delta\theta = T_0 - T_m = 9.85$ K. For small times (the curves closer to $s = 1$), except in the vicinity of the channel inlet $x = 0$, the solidification front is approximately flat (as predicted by Eq. (3.61)). As time progresses, the ice shape for $x > 0.2$ retains the flat shape,

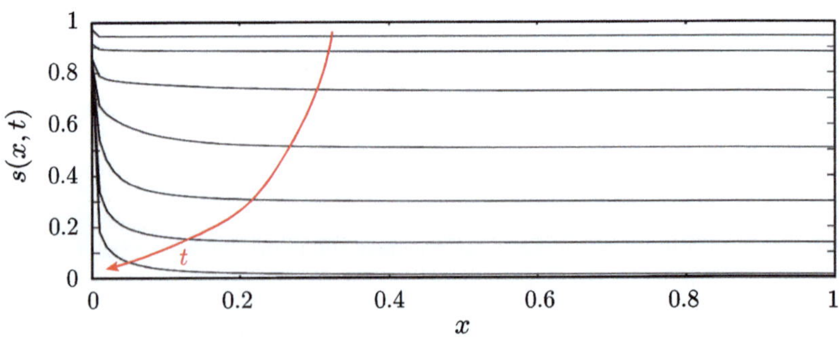

Fig. 3.2 Evolution of the solidification front when $Pe' = K = 1$. The curves follow from the numerical solution of the integro-differential equation (3.59)

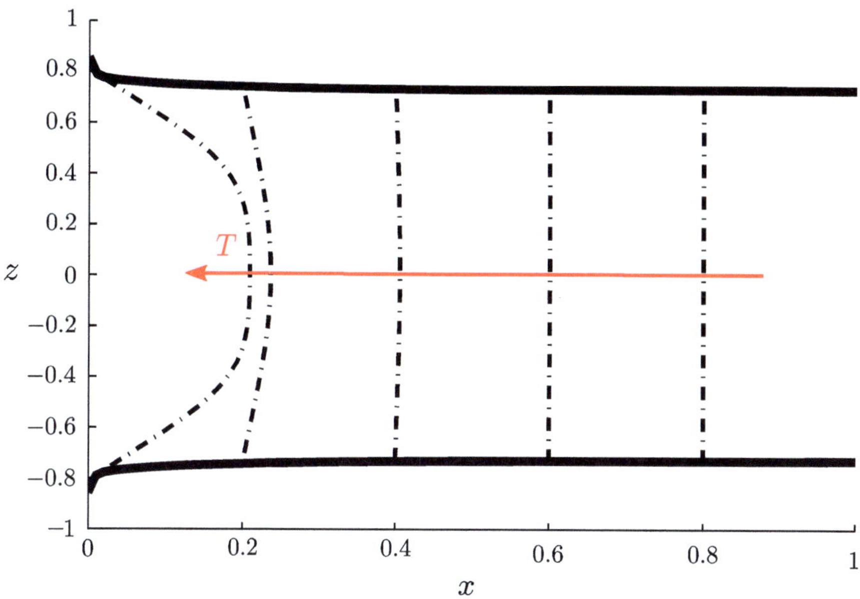

Fig. 3.3 Curves of constant temperature (dash-dot lines) $\theta = [0.0005, 0.0039, 0.0273, 0.1864, 0.9983]$ at time $t = 0.0341$ for the same conditions as in Fig. 3.2, the temperature arrow indicates an increase moving towards the inlet. Thick lines represent the position of the ice-water interface

whereas for smaller x, there is a transition region where the ice shape is curved. The lowest curve corresponds to $t = 0.4754$, and the channel completely closes at $t \approx 0.5$. This corresponds almost exactly with the prediction of Eq. (3.61) where $s(0.5) = 0$. This correspondence is to be expected given that closure will first occur far downstream (large x) and $s \to 0$ consequently $\gamma F \gg 1$ and Eq. (3.61) holds.

Curves of constant fluid temperature, corresponding to the solidification profile shown in Fig. 3.2, at time $t = 0.0341$, are presented in Fig. 3.3. At such an early time, the channel is still wide open and so the flow is relatively fast, allowing heat to penetrate along the channel. The first curve on the left represents the position where $\theta = 0.9983$, i.e. it has barely changed from the inlet value, $T = 1$. As the flow progresses, the temperature becomes approximately constant across the channel while slowly cooling as it moves to the right. At around $x = 0.5$, the temperature has decreased to 1% of its inlet value. The equivalent results at the later time; $t = 0.2379$ are shown in Fig. 3.4. Now the channel is more than half-way closed and the flux significantly reduced. Consequently, the inlet heat does not penetrate far into the channel. The first curve on the left has $\theta \approx 0.28$; the remaining lines have a nondimensional temperature close to zero.

A key result from the study of [109] was that for reduced Peclet numbers below 10, the time for the valve to close was approximately constant $t_{cl} \approx 0.5$. Since the reduced Peclet number $\mathrm{Pe}' = UH^2/\alpha_l L$, this condition is easily achieved in a microchannel where velocities are low and heights small. In fact, this closure time

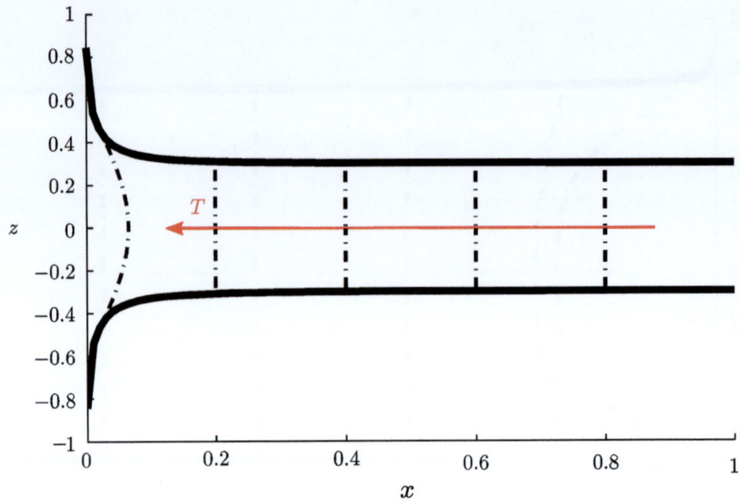

Fig. 3.4 Curves of constant temperature θ (dash-dot lines), at time $t = 0.2379$ for the same conditions as in Fig. 3.2, the temperature arrow indicates an increase moving towards the inlet. Thick lines represent the position of the ice-water interface

may have been predicted from the "small time solution", Eq. (3.61). In both Figs. 3.3 and 3.4, it may be observed that sufficiently far from the inlet θ is independent of z; hence, $\theta_z = 0$. The Stefan condition then reduces to the small time form (3.61). So, far downstream, the solid growth rate satisfies the small time solution and $s \approx 1 - \sqrt{2t}$. Closure first occurs in this region so that $t_{cl} \approx 1/2$.

The dimensional closure time may then be written in terms of the system parameters:

$$t_{cl} = \frac{\rho_s L_m H^2}{2k_s(T_m - T_w)} \quad \text{seconds}\,, \tag{3.62}$$

(recall the time-scale $\tau = k_s \Delta T / \rho_s L_m H^2$). This shows that the closure time increases quadratically with the channel width and decreases linearly as the wall temperature decreases. This constant value for low flow rates was demonstrated experimentally by Chen [34], where for rates below 10 μl/min, the closure time was approximately 6 s. A further observation in [109] was that beyond a certain value, the closure time increases with flow rate and beyond a critical Pe′, the valve will no longer close. This value was shown to vary approximately with the reciprocal of k_l, which permitted the definition of a simple operating criteria. The relation to k_l (and other parameters) may be inferred from Eq. (3.59), where flow dominates for $K = k_l \Delta\theta/(k_s \Delta T) \gg 1$. Put another way, if we divide both sides of Eq. (3.59) for large K, then heat conduction through the solid becomes negligible, and for the equation to balance, we must rescale time such that $\bar{t} = Kt$ represents a long (or slow) time-scale. This time-scale becomes slower as k_l or $\Delta\theta$ increases or k_s, δT decrease.

The case of solidification in a circular pipe is analysed in [109]. The analysis follows in an identical manner, permitting separable and pseudo-steady solutions in the liquid and solid phases, respectively. Conclusions are qualitatively the same as in the Cartesian case (see the exercises at the end of the chapter).

Myers and Low [110] deal with the circular pipe and a fluid with a power law form for the viscosity, using blood and polyethylene oxide as examples. They find that in general shear thinning fluids freeze faster than the Newtonian equivalent under the same pressure drop.

3.3 Contact Melting

Contact melting occurs when a phase change material (PCM) is placed in contact with a surface that is maintained above the melt temperature. The solid begins to melt so that a thin liquid layer develops between the two surfaces. The weight of the solid acts to squeeze out the liquid, and thus the melt layer remains thin.

As a PCM melts, it absorbs energy in the form of latent heat. Conversely, as it freezes, energy is released into the surroundings. These properties have led to PCMs being used in applications such as the transportation of temperature-sensitive materials (e.g. food, pharmaceutical products, and electrical components), thermal energy storage systems in buildings, high-performance textiles, and solar energy-related applications. Practical uses, experiments, models, and analysis of contact melting are discussed in [13, 38, 93].

A schematic of the contact melting process is shown in Fig. 3.5. As the PCM melts, a thin-fluid film forms; this film is squeezed out by the weight of the PCM. The squeezing effect ensures that the film remains thin throughout the process and so the lubrication approximations of Sect. 3.1 hold. In fact, since the flow is in a channel, it bears a strong resemblance to the analysis of the preceding sections with a few key differences, as detailed below.

In the following, we will develop a model for the experiments of Moallemi et al. [93], which involved placing a cuboid PCM, with initial dimension $2L \times 2B \times H_0$, on a flat surface. The top and sides were insulated (this feature is key to the mathematical model). Since heat is gained only through the bottom of the block and, during the melting process, the bottom is everywhere at the constant phase change temperature, the heat transfer occurs only in the z-direction. The liquid layer occupies $z \in (0, h(t))$. The solid height is denoted $H(t)$ where $H(0) = H_0$. The actual height of solid that has melted, $h_m(t)$, differs from the film thickness due to the squeeze film effect $h(t) \neq h_m(t)$. Then the height of the solid over time $H(t) = H_0 - h_m(t)$, while the solid mass, which drives the liquid motion, $M(t) = 4\rho_s LB[H_0 - h_m]$.

The temperatures in the solid and liquid are defined by the reduced system:

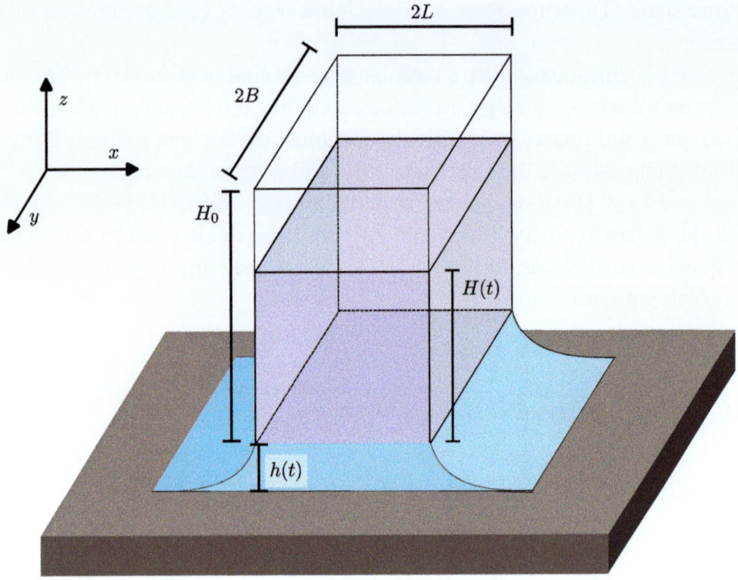

Fig. 3.5 Schematic of contact melting problem. A block of phase change material (purple) rests on a warm substrate (brown); melt liquid (blue) forms between the two solid layers and is squeezed out due to the weight of the PCM

$$\rho_s c_s \frac{\partial T}{\partial t} = k_s \frac{\partial^2 T}{\partial z^2}, \qquad \frac{\partial^2 \theta}{\partial z^2} = 0. \tag{3.63}$$

Here we have anticipated the more formal reduction. The initial height of the solid is of the order of the width; hence, no thin layer reduction is possible in the solid. However, the fluid layer is always thin, suggesting derivatives in the z-direction dominate. Consequently, Eqs. (3.63) are consistent with neglecting terms of order $\mathcal{O}(\varepsilon^2, \varepsilon^2 Pe)$. To ensure that this approximation holds, the size of the reduced Peclet number $\varepsilon^2 Pe$ through a contact melting process is investigated in [115], where it is shown to remain around 0.1 until the very end of the melting process when it peaks below 0.2. Since this is the largest neglected term, it could form the basis of a perturbation expansion with $\varepsilon^2 Pe$ as the small parameter. The studies of [57, 159] both retain the advection term in the liquid heat equation.

The Stefan condition is

$$\rho_s L_m \frac{dh_m}{dt} = k_s \frac{\partial T}{\partial z} - k_l \frac{\partial \theta}{\partial z}\bigg|_{z=h}. \tag{3.64}$$

Applying Newton's second law, $F = ma$, at the liquid-solid interface shows that

$$4\rho_s LB\,[H_0 - h_m(t)]\,g - \int_{-L}^{L}\int_{-B}^{B} p\,dy\,dx = 4\rho_s LB \frac{dV_s}{dt}, \tag{3.65}$$

that is, the difference between the PCM's weight and the pressure exerted by the liquid results in an acceleration, where V_s is the vertical velocity of the solid at the boundary. This velocity may be expressed as the difference between the liquid layer growth rate and the melt rate $V_s = \partial h/\partial t - \partial h_m/\partial t$. Since V_s is of the order of the vertical velocity in the liquid, $w = \varepsilon U$, and we will work on the melting time-scale, then this acceleration term is negligible [78, 115], so that

$$4\rho_s LB\left[H_0 - h_m(t)\right]g = \int_{-L}^{L}\int_{-B}^{B} p\,dy\,dx\,. \tag{3.66}$$

We could of course demonstrate this more formally by writing out the full system of equations here and then neglecting small terms after nondimensionalisation. I hope we are beyond that stage.

To determine the pressure requires knowledge of the flow, which is described by the three-dimensional lubrication approximation:

$$\frac{\partial p}{\partial x} = \mu_l \frac{\partial^2 u}{\partial z^2}\,, \qquad \frac{\partial p}{\partial y} = \mu_l \frac{\partial^2 v}{\partial z^2}\,, \qquad \frac{\partial p}{\partial z} = 0\,, \tag{3.67}$$

$$\frac{\partial u}{\partial x} + \frac{\partial v}{\partial y} + \frac{\partial w}{\partial z} = 0\,. \tag{3.68}$$

Following [115], we may split the contact melting process into three distinct stages. The first stage occurs immediately after the PCM comes into contact with the heated substrate. With a fixed temperature boundary condition, melting would occur instantaneously. In practice, heat transfer is not perfect; hence, initially the PCM is in contact with the substrate for a finite time before melting starts. This is termed the pre-melt stage. It is typically short; here we will define it as lasting for $t \in [0, t_1]$ where t_1 is the time that the temperature at $z = 0$ reaches T_m. The second stage is that in which the liquid layer is first formed and the block begins to float upon it. This occurs for $t \in [t_1, t_2]$ and ends when the temperature at the top of the solid rises (noticeably) above the initial temperature. Finally, the third stage begins at t_2 and ends once all of the block is melted, i.e. $t \in [t_2, t_m]$. Note, the distinction between stages 2 and 3 is purely a requirement of the mathematical model and a different approach may not need this.

In the pre-melt stage, the system of thermal equations must be solved subject to

$$k_s \frac{\partial T}{\partial z} = \mathscr{H}_{ss}(T - T_s)\bigg|_{z=0}\,, \qquad \frac{\partial T}{\partial z}\bigg|_{z=H_0} = 0\,, \qquad T(x, z, 0) = T_0\,. \tag{3.69}$$

The first equation is a Robin condition where $\mathscr{H}_{ss}$ is the heat transfer coefficient between solid and substrate and $T_s > T_m$ is the substrate temperature. The second equation states that the top surface is insulated (as in the experiment), so there is no energy flux. Initially, the PCM is at a constant temperature $T_0 < T_m$. Once melting begins, the conditions are

$$k_l \frac{\partial \theta}{\partial z} = \mathscr{H}_{sl}(\theta - T_s)\Big|_{z=0}, \quad \theta(x, h, t) = T_m\,, \quad T(x, h, t) = T_m\,, \quad \frac{\partial T}{\partial z}\Big|_{z=H+h} = 0\,, \tag{3.70}$$

where $\mathscr{H}_{sl}$ is the substrate-liquid heat transfer coefficient.

No slip at the solid surfaces indicates $u = v = 0$ at $z = 0, h$, while $w = 0$ at $z = 0$. The final condition on the flow, required to determine the liquid vertical velocity at the interface, comes from a mass balance of the form (3.18) with

$$\hat{\mathbf{n}} = (0, 0, 1)\,, \quad \mathbf{u}_s = \left(0, 0, \frac{\partial h}{\partial t} - \frac{\partial h_m}{\partial t}\right), \quad \mathbf{u}_l = (u, v, w)\,, \quad \mathbf{u}_b = \left(0, 0, \frac{\partial h}{\partial t}\right). \tag{3.71}$$

From this, we obtain an expression for the fluid velocity at $z = h$:

$$w|_{z=h} = \frac{dh}{dt} - \frac{\rho_s}{\rho_l}\frac{dh_m}{dt}\,. \tag{3.72}$$

A squeeze film occurs when two nearly parallel surfaces, separated by a thin layer of fluid, move towards or away from each other. They are ubiquitous in the study of gears and bearings and so the theory is well-developed; see [2]. Equation (3.72) is a modification of the boundary condition encountered in a standard squeeze film model. Neglecting the melting component, $h_{mt} = 0$, then the standard form is recovered.

At the outer edge of the block, the fluid encounters the ambient conditions:

$$p|_{x=\pm L} = p|_{y=\pm B} = p_a\,. \tag{3.73}$$

This condition is in fact an approximation. Outside of the block, there will exist a thin film with a free surface (as described in Sect. 3.1.2), which will increase the pressure at the boundary. However, it is assumed that the curvature-induced pressure jump at the free surface is small compared to the pressure due to the weight of the solid and so may be neglected.

We nondimensionalise according to

$$\hat{T} = \frac{T - T_m}{T_s - T_m}, \quad \hat{\theta} = \frac{\theta - T_m}{T_s - T_m}, \quad \hat{x} = \frac{x}{L}, \quad \hat{y} = \frac{y}{B}, \tag{3.74}$$
$$\hat{u} = \frac{u}{U}, \quad \hat{v} = \frac{v}{V}, \quad \hat{t} = \frac{t}{\tau}, \quad \hat{p} = \frac{p - p_a}{P},$$

where $P = \rho_s g H_0$ is the pressure scale (due to the weight of the solid), $U = PH_0^2/(\mu_l L)$, $V = PH_0^2/(\mu_l B)$ are standard velocity scales (recall that for lubrication theory, we previously defined the pressure scale $P = \mu_l U/(\varepsilon^2 L)$). We have chosen the co-ordinate system such that $B > L$, that is, the block is widest in the y direction. In the limit $B \to \infty$, the model reduces to the two-dimensional form studied by Myers, Mitchell, and Muchatibaya [115]. For the nondimensional height, we will use two distinct scales: in the solid, we scale with the initial height H_0; in the liquid, we use a representative small height-scale, H' and hence write

$$\hat{z} = \frac{z}{H'}, \qquad \bar{z} = \frac{z}{H_0}, \qquad \hat{w} = \frac{w}{W}, \tag{3.75}$$

where $W = H'U/L$ (take care here, we have a lot of H's lying around). The two height-scales result in two z co-ordinates that are related by $\bar{z} = \varepsilon_1 \hat{z}$ where $\varepsilon_1 = H'/H_0 \ll 1$. The small size of ε_1 suggests a possible simplification of the system; however, here we will retain all terms of this size since a solution is possible without further reduction. Dropping the hats, but retaining the overbar to distinguish between the z co-ordinates, gives

$$A_1 \frac{\partial T}{\partial t} = \frac{\partial^2 T}{\partial \bar{z}^2}, \qquad \frac{\partial^2 \theta}{\partial z^2} = 0, \tag{3.76}$$

where $A_1 = \rho_s c_s H_0^2/(k_s \tau)$ (having scaled h_m with H_0). The Stefan condition identifies the melting time-scale $\tau = \rho_s L_m H_0 H'/(k_l \Delta T)$:

$$\frac{dh_m}{dt} = A_2 \frac{\partial T}{\partial \bar{z}}\bigg|_{\bar{z}=\varepsilon_1 h} - \frac{\partial \theta}{\partial z}\bigg|_{z=h}, \tag{3.77}$$

where $A_2 = \varepsilon_1 k_s/k_l$. The presence of the small parameter here indicates that melting is driven by heat flow through the liquid and, beyond the initial stage, that heat flow in the solid plays a small role. The lubrication equations and force balance become

$$\frac{\partial p}{\partial x} = \frac{\partial^2 u}{\partial z^2}, \qquad \frac{\partial p}{\partial y} = \frac{\partial^2 v}{\partial z^2}, \qquad \frac{\partial p}{\partial z} = 0, \tag{3.78}$$

$$\frac{\partial u}{\partial x} + \lambda^2 \frac{\partial v}{\partial y} + \frac{\partial w}{\partial z} = 0, \tag{3.79}$$

$$4\left[1 - h_m(t)\right] = \int_{-1}^{1} \int_{-1}^{1} p \, dy \, dx, \tag{3.80}$$

where $\lambda = L/B$ arises due to the different scaling of velocity in the x and y directions. The two-dimensional model may then be retrieved in the limit $\lambda \to 0$.

In the pre-melt stage, there is no fluid, so the only conditions required are

$$\frac{\partial T}{\partial \bar{z}} = Bi(T-1)\bigg|_{\bar{z}=0}, \qquad \frac{\partial T}{\partial \bar{z}}\bigg|_{\bar{z}=1} = 0, \qquad T(x, \bar{z}, 0) = T_0, \tag{3.81}$$

where the Biot number $Bi = \mathscr{H}_{ss} H_0/k_s$ (denoting the relative importance of heat transfer at the boundary to conduction) and $T_0 < 0$. Once melting commences, the fluid layer must be accounted for:

$$\left.\frac{\partial \theta}{\partial z} = Nu(\theta - 1)\right|_{z=0}, \qquad \theta(x, z = h, t) = 0, \tag{3.82}$$

$$T(x, \bar{z} = \varepsilon_1 h, t) = 0, \qquad \left.\frac{\partial T}{\partial \bar{z}}\right|_{\bar{z}=H+\varepsilon_1 h} = 0, \tag{3.83}$$

where the Nusselt number $Nu = \mathscr{H}_{sl} H'/k_l$ (the difference between the Nusselt and Biot numbers is that the first takes the liquid conductivity, while the latter the solid conductivity). At the substrate $u = v = w = 0$, at the solid-liquid interface $u = v = 0$, and

$$w|_{z=h} = \frac{L}{U\tau}\frac{dh}{dt} - A_3\frac{dh_m}{dt}, \tag{3.84}$$

where $A_3 = \rho_s H_0 L/(\rho_l H' U\tau)$. The pressure is ambient at the edge of the block:

$$p|_{x=\pm 1} = p|_{y=\pm 1} = 0. \tag{3.85}$$

According to the final equation of (3.78), at least to leading order in the lubrication approximation, $p_z = 0$ so that $p = p(x, y, t)$. This permits the integration of the first two equations in (3.78) to determine the in-plane fluid velocity:

$$u = \frac{p_x}{2} z(z - h) \qquad v = \frac{p_y}{2} z(z - h). \tag{3.86}$$

The incompressibility condition (3.79) may be integrated across the film:

$$w|_{z=h} - w|_{z=0} = -\int_0^h \frac{\partial u}{\partial x} + \lambda^2 \frac{\partial v}{\partial y}\, dz. \tag{3.87}$$

Substituting for the various velocities, via (3.84, 3.86), and carrying out the integration leads to

$$\frac{L}{U\tau}\frac{dh}{dt} - A_3\frac{dh_m}{dt} = \frac{h^3}{12}\nabla^2_{xy} p, \tag{3.88}$$

where $\nabla^2_{xy} = \partial_{xx} + \lambda^2 \partial_{yy}$ is a modified Laplacian operator. Since h, h_m are functions of time alone, this may be rewritten as a form of Poisson's equation:

$$\nabla^2_{xy} p = f(t), \qquad f(t) = \frac{12}{h^3}\left[\frac{L}{U\tau}\frac{dh}{dt} - A_3\frac{dh_m}{dt}\right], \tag{3.89}$$

which may be solved via an eigenfunction expansion of the form

$$p = \sum_{n=1}^{\infty} \Phi_n(Y,t)\sin(\omega_n X)\,, \tag{3.90}$$

where the shifted co-ordinates are $X = (x+1)/2$, $Y = (y+1)/2$, and $\omega_n = n\pi$. Substituting back into Poisson's equation,

$$\sum_{n=1}^{\infty}\left(\lambda^2\frac{\partial^2\Phi_n}{\partial Y^2} - \omega_n^2\Phi_n\right)\sin(\omega_n X) = 4f\,. \tag{3.91}$$

Using the orthogonality of the eigenfunctions,

$$\frac{\partial^2\Phi_n}{\partial Y^2} - \frac{\omega_n^2}{\lambda^2}\Phi_n = \frac{8f(t)}{\lambda^2}\int_0^1 \sin(\omega_n X)\,dX = \frac{8f(t)(1-(-1)^n)}{\omega_n\lambda^2} = F_n(t)\,. \tag{3.92}$$

Applying $\Phi_n = 0$ at $Y = 0, 1$ determines

$$\Phi_n = \frac{\lambda^2 F_n}{\omega_n^2}\left[\frac{(e^{\omega_n/\lambda}-1)}{e^{\omega_n/\lambda}-e^{-\omega_n/\lambda}}e^{-\omega_n Y/\lambda} + \frac{(1-e^{-\omega_n/\lambda})}{e^{\omega_n/\lambda}-e^{-\omega_n/\lambda}}e^{\omega_n Y/\lambda} - 1\right]. \tag{3.93}$$

The force balance (3.80) may be written as

$$1-h_m(t) = \int_0^1\int_0^1 p(X,Y,t)\,dY\,dX = \sum_{n=1}^{\infty}\int_0^1\int_0^1 \Phi_n(Y,t)\sin(\omega_n X)\,dY\,dX\,. \tag{3.94}$$

Series solutions of the form (3.94) often decay rapidly with increasing n; here the decrease is of the order $1/\omega_n^5 = 1/(n\pi)^5$. In [78], only the first term of the series was employed, and further it was assumed that $\exp(\omega_1/\lambda) \gg 1$ (this may be more convincing after noting that $e^{\pi} \approx 23.1$, $e^{-\pi} \approx 0.04$ and our system is such that $\lambda = L/B \leq 1$). From this, a simple equation of the form

$$1 - h_m \approx \Lambda f\,, \tag{3.95}$$

may be obtained (the form of Λ will be derived in the exercises at the end of the chapter). For the case studied in [78], where $\lambda \approx 1/6$, it was shown that there was no noticeable difference in results when using only the first term and the first 50. Consequently, we may write equation (3.88) as

$$\frac{L}{U\tau}\frac{dh}{dt} - A_3\frac{dh_m}{dt} = \frac{h^3}{12\Lambda}(1-h_m)\,. \tag{3.96}$$

This equation relates the two unknown heights, that of the liquid film and the amount of melted liquid. To close the system, we look to the Stefan condition (3.77) and so must first examine the temperature profiles.

3.3.1 Stage 1: Pre-melt Stage

This is a very short stage; in [78], it is shown to occupy only $10^{-4}\%$ of the total process time for the experiments of [93]. However, it is useful in that it determines the temperature profile in the PCM once melting starts. This stage may be recognised as a small time solution that provides the initial condition for the subsequent melting stage. In [78], the pre-melt stage is analysed in two ways, using a Laplace transform (which may be found in [29]) and the TIM. This provides a check on the accuracy of the TIM result: specifically, it is shown that the difference in times when Stage 1 ends as predicted by the two methods are within 2%. Since the TIM is employed in the subsequent stages, for consistency, we will follow this approach now.

To apply the heat balance methods of Sect. 2.2.4, we first replace the boundary condition at the top of the solid block, $T_z|_{\bar{z}=1}$ (see Eqs. (3.81)) with two conditions that define the heat penetration depth $T|_{\bar{z}=\delta} = T_0$, where $\delta \leq 1$ and, for a smooth transition to the constant value, $T_{\bar{z}}|_{\bar{z}=\delta} = 0$. The heat balance form is then

$$T = a_0 + a_1\left(1 - \frac{\bar{z}}{\delta}\right) + a_n\left(1 - \frac{\bar{z}}{\delta}\right)^n, \tag{3.97}$$

where $n \geq 1$. Subject to the boundary conditions at $\bar{z} = \delta$, this reduces to

$$T = T_0 + a_n\left(1 - \frac{\bar{z}}{\delta}\right)^n. \tag{3.98}$$

The coefficient a_n is determined from the condition at $\bar{z} = 0$ giving

$$T = T_0 + \frac{\delta Bi(1 - T_0)}{n + \delta Bi}\left(1 - \frac{\bar{z}}{\delta}\right)^n; \tag{3.99}$$

see Eq. (2.114). Since we have applied a Robin condition, the minimum error via the TIM technique is obtained with $n = 3.584$; see Sect. 2.2.4.1.

The pre-melt stage ends when the temperature at $\bar{z} = 0$ reaches the melt temperature, $T(0, t) = 0$. From Eq. (3.99), this determines the penetration depth:

$$\delta(t_1) = \delta_1 = -\frac{nT_0}{Bi}. \tag{3.100}$$

However, knowing the depth does not determine the time at which this stage ends, t_1. For this, we need to evaluate the heat balance integral using the temperature (3.99). This leads to a first-order ordinary differential equation for $\delta(t)$ with implicit solution:

$$t = \frac{A_1}{n(n+1)}\left[\frac{\delta^2}{2} + \frac{n\delta}{Bi} - \frac{n^2}{Bi^2}\ln\left(\frac{n+\delta Bi}{n}\right)\right]; \tag{3.101}$$

see Eq. (2.114). Substituting for $\delta = \delta_1$ then provides an expression for t_1 for given values of the system parameters $Bi = \mathscr{H}_{ss} H_0/k_s$, $A_1 = \rho_s c_s H_0^2/(k_s \tau)$. After going to the trouble of calculating this, we note that in general, t_1 is so small that it may be neglected. As an example, consider the pre-melt stage examined in [78]; this lasts for approximately 10^{-2} s compared to a total melt time of 1525 s. Consequently, we may start Stage 2 at $t = 0$, with the condition $\delta(0) = \delta_1$ and wonder why we even wrote down Eq. (3.101).

To verify the accuracy of the above solution, we note that Carslaw and Jaeger [29, §2.7] derive a solution valid for a semi-infinite media, subject to a Robin condition at $z = 0$:

$$T = 1 + (T_0 - 1)\left[\operatorname{erf}\left(z\sqrt{\frac{A_1}{4t}}\right) + \exp\left(Bi\, z + t\, Bi^2/A_1\right)\right.$$
$$\left. \times \operatorname{erfc}\left(z\sqrt{\frac{A_1}{4t}} + Bi\sqrt{\frac{t}{A_1}}\right)\right]. \tag{3.102}$$

Provided the temperature remains negligible near the top of the block, this will provide an accurate solution (and recall this solution only holds to very small times). In Fig. 3.6, we compare the temperature profiles predicted by the TIM and Eq. (3.102) at the end of Stage 1, for parameter values appropriate to the experiments of [93]. A full list of parameter values are shown in Table 3.1. The position δ_1 is marked by an asterisk – this indicates how heat balance methods define a sufficiently large domain to capture significant temperature changes. Overall, it is clear that the agreement is excellent, confirming the accuracy of the TIM and so confirming that it may be used in the following stages, where no analytical solution is available.

3.3.2 Stage 2: Heating Throughout the Solid

In Stage 2, there is an evolving liquid layer, so the thermal problem is now described by the two heat equations (3.76) subject to (3.82, 3.83). The integration of the liquid heat equation of (3.76) is straightforward, leading to

$$\theta = \frac{Nu}{1 + hNu}(h - z). \tag{3.103}$$

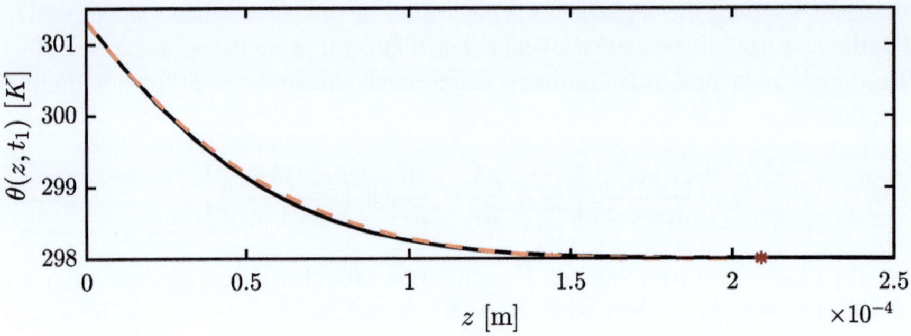

Fig. 3.6 Comparison of temperature profiles at the end of Stage 1 predicted by the TIM (3.99) (dashed orange line) and similarity solution (solid line). The position δ_1 is marked with a red $*$. Parameter values are provided in Table 3.1. Note, to clarify the physical problem, we present results in dimensional form, although the results were obtained through a nondimensional analysis

Table 3.1 Typical parameter values for solid/liquid n-octadecane, taken from [78, 93, 115]

Symbol	Value	Units
L, B, H_0	0.0175, 0.115, 0.055	m
ρ_s, ρ_l	930, 771.2	kg m^{-3}
c_s, c_l	2151, 2161	m^2 s^{-2} K^{-1}
k_s, k_l	0.3723, 0.15	W m^{-1} K^{-1}
μ_l	0.004	N s m^{-2}
L_m	2.435×10^5	J kg^{-1}
T_s, T_m, T_0	308.2, 301.33, 298	K
$\mathscr{H}_{sl}$	5740	W m^{-2} K^{-1}

In the solid, applying $T(\delta, t) = T_0$, $T_{\bar{z}}(\delta, t) = 0$, and $T(\varepsilon_1 h, t) = 0$ to the heat balance temperature profile of (3.97) results in

$$T = T_0 \left[1 - \frac{(\delta - \bar{z})^n}{(\delta - \varepsilon_1 h)^n} \right], \tag{3.104}$$

where, since the temperature is fixed at the lower boundary, the TIM imposes $n = 2.233$. Applying the heat balance integral to the solid heat equation of (3.76),

$$\int_{\varepsilon_1 h}^{\delta} A_1 \frac{\partial T}{\partial t} \, d\bar{z} = \int_{\varepsilon_1 h}^{\delta} \frac{\partial^2 T}{\partial \bar{z}^2} \, d\bar{z} = - \left. \frac{\partial T}{\partial \bar{z}} \right|_{\bar{z} = \varepsilon_1 h} . \tag{3.105}$$

For the final equality, we have applied the condition on the gradient at $\bar{z} = \delta$. Applying Leibniz' theorem to the integral,

$$\frac{d}{dt} \int_{\varepsilon_1 h}^{\delta} T \, d\bar{z} = \int_{\varepsilon_1 h}^{\delta} \frac{\partial T}{\partial t} \, d\bar{z} + \frac{d\delta}{dt} T(\delta, t) - \varepsilon_1 \frac{dh}{dt} T(\varepsilon_1 h, t) , \tag{3.106}$$

and so

$$A_1 \int_{\varepsilon_1 h}^{\delta} \frac{\partial T}{\partial t}\, d\bar{z} = A_1 \left[\frac{d}{dt} \int_{\varepsilon_1 h}^{\delta} T\, d\bar{z} - T_0 \frac{d\delta}{dt} \right] . \tag{3.107}$$

Evaluating all terms in the above heat balance integral, with the temperature expression (3.104), we obtain

$$A_1 \frac{d}{dt}(\delta + n\varepsilon_1 h) = \frac{n(n+1)}{\delta - \varepsilon_1 h} . \tag{3.108}$$

This equation links δ, h. The Stefan condition (3.77) provides a further relation:

$$\frac{dh_m}{dt} = A_2 \frac{nT_0}{\delta - \varepsilon_1 h} + \frac{Nu}{1 + hNu} . \tag{3.109}$$

These two first-order differential equations contain three unknown functions, h, h_m, δ. Together with the force balance (3.96) and the conditions $h(0) = h_m(0) = 0$, $\delta(0) = \delta_1$ (as determined in Stage 1), the system is closed. Stage 2 ends at time $t = t_2$ when the temperature starts to rise at the top of the block. So t_2 is defined as the time when $\delta = \delta_2 = H + \varepsilon_1 h$.

3.3.3 Stage 3: Final Melting

Physically, there is no change between Stages 2 and 3; this final stage is really required due to the form of the integral method. Until now, the integral method has been subject to $T(\delta, t) = T_0$. If the temperature in the PCM is everywhere above this value, then the condition no longer holds. So now we define the temperature at the top of the block $T(H + \varepsilon_1 h, t) = T_H(t)$, where T_H is unknown. Further, since the top is insulated, $T_{\bar{z}} = 0$ there. Since the heat has already penetrated all the way through the block, $\delta(t)$ can no longer be considered a system parameter; instead, we now have to determine $T_H(t)$.

Writing the solid temperature as a polynomial in $(H + \varepsilon h - z)$ leads to

$$T = T_H \left(1 - \frac{(H + \varepsilon h - \bar{z})^n}{H^n} \right) , \tag{3.110}$$

where again $n = 2.233$. Carrying through a standard heat balance integral would lead to a complicated expression for T_H. To simplify this type of problem, Goodman [56] proposed a function, Φ, defined as the integral of the temperature over the domain of interest:

$$\Phi = T_H \int_{\varepsilon h}^{H+\varepsilon h} \left(1 - \frac{(H + \varepsilon h - \bar{z})^n}{H^n} \right) d\bar{z} = \frac{nHT_H}{n+1} . \tag{3.111}$$

This determines T_H in terms of Φ. Integrating the heat equation through the solid region gives

$$A_1\left[\frac{d\Phi}{dt} - T_H\frac{d}{dt}(H+\varepsilon h)\right] = -\frac{nT_H}{H}\,. \tag{3.112}$$

Substituting for T_H from Eq. (3.111) and recalling that $H = 1 - h_m(t)$, Eq. (3.112) provides a relation between Φ, h, h_m. The fluid temperature is the same as in Stage 2; hence, the Stefan condition (3.77) may be written as

$$\frac{dh_m}{dt} = A_2\frac{nT_H}{1-h_m} + \frac{Nu}{1+hNu}\,. \tag{3.113}$$

Again we close the system via the force balance (3.96) and the conditions determined at the end of Stage 2, $h(t_2) = h_2$, $h_m(t_2) = h_{m2}$, $T_H(t_2) = T_0$, and hence $\Phi(t_2) = n(1-h_{m2})T_0/(n+1)$. The end of Stage 3 is the end of the melt process, when all the solid has disappeared, that is, $H = 0$ and $h_m = 1$.

3.3.4 Results

The analysis so far has demonstrated how the phase change of a three-dimensional block lying on a thin fluid layer may be analysed using lubrication theory and the TIM. For the analysis, the process must be split into three stages. The first dealing with the thermal evolution in the initial solid block, to determine when melting begins. The next two stages involve applying the lubrication equations and the TIM, resulting in a system of three first-order differential equations in each stage. These are easily solved using standard numerical techniques, and, to be clear, solving three first-order ordinary differential equations is significantly simpler than solving the original system of flow and heat equations with two moving boundaries.

For a theory to be convincing, it must be well justified and somehow verified. In this case, we demonstrate the validity through comparison with the experimental data of [93]. Their study involved placing a cuboid of solid n-octadecane, at an initial temperature of 298 K, on a substrate maintained at 308.2 K. The top and sides were insulated, such that the heat flow was primarily from the substrate. N-Octadecane is a common phase change material with the useful property that it melts at a pleasant 301.33 K (around 28 °C). The appropriate parameter values for the experiment are presented in Table 3.1.

The evolution of the dimensional height of the n-octadecane block is shown in Fig. 3.7. The solid line represents the solution described in the previous sections, and circles are the experimental data points. The model contains a single unknown, the solid-liquid heat transfer coefficient. This parameter has been adjusted to obtain the result shown in the figure.

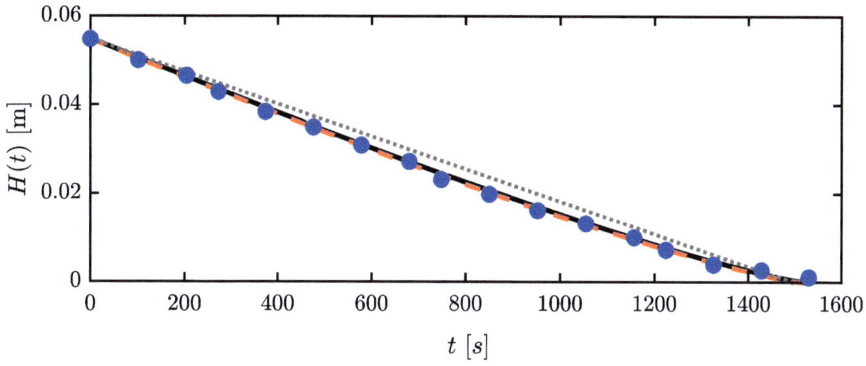

Fig. 3.7 Dimensional evolution of the thickness of an n-octadecane block. Blue circles represent the experimental data, and the solid black line represents the full solution developed in Sect. 3.3.1–3.3.3, the dotted grey line the linear approximation $h_m = Ct$ and the dashed red line (hardly distinguishable from the solid line) the solution of (3.115)

To obtain such a close fit, we have chosen a value for the heat transfer coefficient $\mathscr{H}_{sl}$ = 5740 W/(m^{2}K). Note, this is higher than the value quoted in [78], where the incorrect block width was used (the change in width changes the calculation for the weight of the block that affects the thickness of the film, which in turn affects the heat transfer calculation). The good agreement between the model and experiment gives confidence in the model: an incorrect model with a single free parameter is unlikely to match the 17 data points plotted in the figure. This issue is discussed in detail when comparing a similar model to experimental data in [98].

In [78], two simplifications of the above model are considered. The first is a simple linear approximation motivated by the observation that the decrease in H shown in Fig. 3.7 is approximately linear and hence the increase in h_m is also linear in time $h_m = Ct$. Assuming C to be constant, it is sufficient to find its value at any time and then we will have an explicit expression for h_m. At the beginning of Stage 2, we know that $h = 0, \delta = \delta_1$ and so the Stefan condition (3.109) may be written as

$$\frac{dh_m}{dt} = C = A_2 \frac{nT_0}{\delta_1} + Nu = Nu - A_1 Bi \,, \tag{3.114}$$

after applying the definition of δ_1 of Eq. (3.100). Recalling the various definitions, we may write $A_1 = (k_l/k_s) \cdot 1/(\varepsilon_1 \beta)$, $A_2 = \varepsilon_1 k_s/k_l$, and H_0 is the original thickness of the block, H' is the liquid height-scale, β is the Stefan number, and $\varepsilon_1 = H'/H_0 \ll 1$. The dotted line of Fig. 3.7 comes from this approximation. This provides a very rough estimate for the evolution of the block. The second method involves noting that, except for at very small times, the temperature gradient in the solid adjacent to the melt front is significantly lower than that in the liquid. If the solid temperature is neglected, then the Stefan conditions in Stages 2 and 3 coincide, while the mass balance is unaffected by the change. The result is a two-equation system for $h(t), h_m(t)$:

$$\frac{dh_m}{dt} = \frac{Nu}{1 + hNu}, \qquad \frac{L}{U\tau}\frac{dh}{dt} - A_3\frac{dh_m}{dt} = \frac{h^3}{12\Lambda}(1 - h_m), \tag{3.115}$$

which holds throughout the melting process, although it misses the initial (and very short) heating stage. These equations are to be solved subject to $h(0) = h_m(0) = 0$. While this system also requires a numerical solution, it has the advantage of avoiding the three-stage process, and as may be seen from the red dashed line of Fig. 3.7, it provides excellent agreement with the experimental data for all time. It was concluded that, at least for the contact melting of n-octadecane, the two-equation model provided a simpler and sufficiently accurate description of the phase change process.

3.3.5 Related Problems

A nanofluid consists of a suspension of nanoparticles in a base fluid. There have been numerous remarkable claims concerning their enhanced thermal properties. However, there is a body of experimental evidence, involving a wide variety of notable institutes, which have called these claims into questions; see [20, 21]. Similarly, a plethora of mathematical results supporting these claims have been demonstrated as erroneous; see Myers et al. [117]. Inspired by various experiments concerning the use of nanofluid-based PCMs, the methodology of the preceding sections (but with parameters appropriate to nanofluids) was employed by MacDevette and Myers [79]. In keeping with the conclusions of [117], they demonstrate that the nanofluid PCM is not necessarily more efficient at releasing energy than the base fluid and that the efficiency is dependent on system size. They conclude that the addition of nanoparticles to a water-based phase change material results in only small differences in energy storage (and not always positive), suggesting that nanofluid PCMs are not a practical option.

Cregan et al. [38] generalise the above work to include temperature-dependent thermal conductivity and viscosity. One motivating factor for this work was to see if a nanofluid could be tuned to improve the process. They included temperature dependence into the model of [79] applied to an n-octadecane-based fluid and found results similar to the constant property case. It was shown that this was due to competition between the decrease in viscosity and conductivity as the temperature increased. They suggested that at least in the case of n-octadecane, the accuracy of the constant property result when compared to the more realistic variable property case is somewhat fortuitous since when taken separately, the changes in viscosity and conductivity lead to a significant variation in melt times, but when taken together, they act to cancel each other out.

The Leidenfrost effect is a well-known phenomenon, observed when hot water is dropped on to a surface that is at a temperature well above the evaporation temperature. The droplet impacts on the substrate, then the region in contact with the surface vaporizes, and subsequently the droplet floats on a layer of its own vapor. As well as being a common sight in the kitchen, this phenomenon has important technical applications in processes such as spray cooling, combustion of fuel, and cooling systems for nuclear reactors. Myers and Charpin [98] use a similar analysis to that described in Sect. 3.3 to study this process. Their work was simplified by the assumption that convection occurs in the liquid drop, so permitting mixing and an approximately constant temperature throughout the liquid. On the other hand, the model must be linked to a dynamic Young-Laplace equation to determine the droplet shape, which complicates the calculations. Matching with experimental data demonstrated that evaporation must occur all around the droplet, not just in the area immediately above the substrate.

Bejan [13, §10.3] considers the melting of a PCM, initially at the phase change temperature, which is moving at a constant horizontal speed over the substrate. This scenario is easily reproduced by the above analysis, simply by setting the tangential boundary condition on the fluid velocity to $u|_{z=h} = U_h$, where U_h is the PCM velocity. The extension of the theory outlined in the previous section to that of a moving substrate is detailed in MacDevette et al. [78].

3.4 Atmospheric Icing

Ice accretion on surfaces is a source of concern for a number of industries. Power cables and radio masts have been damaged or destroyed on numerous occasions due to the added weight of the ice or an increase in aerodynamic interaction, leading to unacceptable movement such as galloping [129]. During the 1998 North American Ice Storm as well as widespread devastation of trees, over 1000 transmission towers collapsed, resulting in power outages lasting between days and months throughout the northern USA and Canada. It is estimated to have resulted in between 5 and 7 billion US$ [155]. On aircraft various in-flight incidents and crashes have been shown to be directly attributable to ice accretion. This may be due to ice buildup on an aerofoil leading edge; for example, an accretion of only 0.4 mm on a wing upper surface may reduce lift by 25% and decrease the stall angle of attack by 6°[67]. Alternatively, ice may build up on a noncritical surface and subsequently break off, damaging downstream components. For these reasons, most commercial and military aircraft are fitted with anti- and de-icing equipment. However, these devices require design compromises; they also use a considerable amount of energy; a typical system for a long-range passenger jet consumes of the order 10^5 W when active.

Ice growth on aircraft and structures is primarily due to the impact of supercooled water droplets on a cold surface. Two basic types of ice may form: rime or glaze. Rime ice is associated with small droplets and very cold conditions, typically when the ambient temperature is below −15 °C, so the impacting droplets freeze almost instantaneously. In milder conditions or when the water droplets are large, glaze ice forms. In this case, only a fraction of the droplet will freeze instantaneously; the rest can remain liquid for some time and may flow as "runback" water.

In the following, we will discuss ice accretion in the context of aircraft icing; however, the models are applicable to ice growth on any cold structure. In particular, ice accretion on a wind turbine occurs in an identical manner to that on aircraft. Ice buildup on power lines and related equipment also occurs as a result of in-cloud icing or freezing rain and drizzle. In a wider context, solidification from a flowing liquid or a droplet spray is of interest in the casting of metals and spray forming, lava flows, and hydrate buildup in oil pipelines [113, 114].

A typical ice accretion begins with supercooled water droplets impacting on a cold surface. If both droplets and surface are below the freezing temperature, the initial ice growth is dry; this is termed rime ice. As the accretion grows, the kinetic energy of the incoming droplets and the release of latent heat act to increase the temperature. Under certain conditions, typically in the range $[-15, 0]$°C, after a sufficient time, a portion of the incoming liquid remains as a liquid and is then free to flow over the existing ice; it may subsequently freeze to form glaze ice. The difference in formation processes means that rime and glaze are easily distinguished. Rime contains air pockets, which makes it opaque; with glaze the flowing water can fill the air pockets, resulting in a transparent layer; hence, glaze is also commonly known as clear ice. Glaze is significantly denser than rime ice.

Since rime ice forms with no water motion, it is easily predicted via CFD packages that may track the air and water droplet flow around the surface: rime ice forms exactly where the droplets impact the surface. Glaze ice is much more difficult to predict, since the flowing water layer results in ice extending far beyond the impact zone and unexpected ice shapes.

The standard model for ice growth from incoming droplets used in early aircraft icing codes is termed the Messinger model [85]. This is a one-dimensional equilibrium energy balance. In Myers and Myers and Hammond [97, 108], it is shown that the Messinger model must always predict a slower growth rate than in reality. Employing the energy sources of the Messinger model in a Stefan problem, a time-dependent model was developed in [97]. The Messinger model appears as a particular limit of this formulation. In a series of papers, Myers and co-workers [16, 104, 113, 114] extend this model and apply it to a variety of dynamic situations. It is now referred to as the Myers model (see [28]) (this was not my idea! But it is a catchy name) or the Extended Messinger Model (less catchy, I suggest sticking to the former) and is employed in commercial aircraft icing software.

Without any further ado, we will now leap straight into the mathematical model, with a few sensible (and tested) assumptions. The two-dimensional model configuration is shown in Fig. 3.8. It is assumed that the ice and water layers remain thin such that the basic one-dimensional (nondimensional) model of [97, 108] holds:

$$\frac{\partial^2 T}{\partial z^2} = \frac{\partial^2 \theta}{\partial z^2} = 0 \,. \tag{3.116}$$

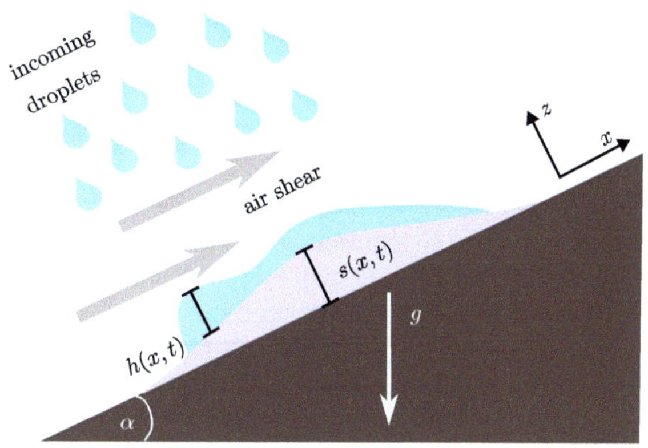

Fig. 3.8 Configuration for two-dimensional ice accretion due to water droplets impacting on a solid surface maintained below the freezing temperature. The ice occupies $z \in [0, s(x,t)]$; the fluid layer occupies $z \in [s(x,t), s(x,t) + h(x,t)]$

In aircraft icing situations, this has been shown to be valid for ice accretions below 2.4 cm and water films less than 3 mm thick [108]. Experiments and numerical studies indicate a water thickness of the order 0.1 mm, while aviation guidelines prohibit the buildup of thick ice layers, so these restrictions are usually satisfied.

The pseudo-steady equations are subject to the following conditions. For rime ice growth,

$$T(x,0,t) = -1\,, \qquad \left.\frac{\partial T}{\partial z}\right|_{z=s} = E_{0r} - E_{1r}\, T|_{z=s}\,, \tag{3.117}$$

where the coefficients E_{ir} encompass all the energy sources and sinks. The heat sinks include convective heat transfer, sublimative heat loss, and cooling by incoming droplets. The heat sources are kinetic energy of incoming droplets, release of latent heat, and aerodynamic heating. Here, purely to make the following analysis clearer, we follow the work of [114], assuming perfect thermal contact at the substrate and that the metal substrate has a much greater thermal mass than the ice and conducts heat rapidly, so permitting a fixed boundary temperature. Subsequent authors have considered different boundary conditions and three-dimensional growth.

For glaze ice growth, appropriate boundary conditions are

$$T(x,0,t) = -1\,, \qquad T(x,s,t) = \theta(x,s,t) = 0\,, \tag{3.118}$$

$$\left.\frac{\partial \theta}{\partial z}\right|_{z=s+h} = E_{0g} - E_{1g}\, \theta|_{z=s+h}\,. \tag{3.119}$$

The energy terms take a different value from the rime case since the convective cooling here occurs over a liquid surface, sublimation is replaced by evaporation, and latent heat release must also be included in E_{0r}. At the ice-water interface, the temperature is at the phase change temperature.

Growth occurs in two stages. The first stage is rime ice growth. In this case, there is no water layer, $h = 0$, so the mass balance, Eq. (3.32), reduces to

$$\frac{\dot{m}}{\rho_l} = \frac{\rho_s}{\rho_l}\frac{L}{U\tau_f}\frac{\partial s}{\partial t}, \tag{3.120}$$

where $\tau_f = L/U$ (conveniently the flow time-scale drops out of the static case). The incoming mass rate may vary with space and time: the air flow and hence rate of droplets impacting the surface may vary as the ice grows on the substrate. In this case, the above mass balance must be integrated numerically. However, for sufficiently small time periods, $\dot{m}$ may be assumed to be approximately constant; applying $s(x, y, 0) = 0$ then leads to

$$s = \frac{\dot{m}}{\rho_s}t \tag{3.121}$$

and the rime ice shape is therefore determined by the form of the incoming mass. Note, the approximation of constant $\dot{m}$ is exploited in "multistep" icing calculations, where the growth is calculated for a short time period with fixed ambient conditions. Air flow and droplet trajectories are then recalculated according to the new aerofoil plus ice shape and another icing calculation carried out with the new parameters. This method avoids computationally expensive continuous recalculation of meshes, flow, and trajectories.

The rime period ends when (and if) the ice temperature ever rises above the freezing point. Integrating the pseudo-steady heat equation for T and applying the rime boundary conditions gives the temperature

$$T = \frac{E_{0r} + E_{1r}}{1 + E_{1r}s}z - 1\,. \tag{3.122}$$

Since the temperature $T(x, 0, t) = -1$ is fixed at the substrate, the temperature takes its greatest value at the surface $z = s(x, t)$. Setting $T(x, s, t) = 0$ leads to

$$s = \frac{1}{E_{0r}} \quad \Rightarrow \quad t = \frac{\rho_s}{E_{0r}\dot{m}}\,. \tag{3.123}$$

Interestingly, this relation indicates that the thickness at which fluid first appears (and consequently the time) depends only on the heat source terms. This could have been inferred from the boundary condition (3.117), where the surface heat flux $T_z \to E_{0r}$ as the surface temperature $T(x, s, t) \to 0$. This simple model for rime ice growth was tested in [108]. Figure 3.9 shows a thin section of a piece of ice grown in an icing wind tunnel. The microstructure is highlighted by the use of a strong background light source, which may be partially seen to the right of the ice section. The accretion was formed in approximately 12 minutes on a standard aluminium aerofoil (a NACA 9901, for those with a real interest) with dimensions of 40 ×

Fig. 3.9 Microstructure of a piece of ice grown on a model aluminium aerofoil NACA 9901 at $-10\,^\circ$C; see [108]

4.8 cm; hence, the curvature is considerably greater than that on an aircraft wing. The ambient freestream and initial substrate temperature was approximately -10°C, the freestream velocity was 90 m/s, and the liquid water concentration was 0.1 g/m^3. The ice grew from right to left into the air flow. A layer of fine columnar grains, associated with rime ice growth, can be seen near the original growth front. As the accretion has continued to grow, the grain structure has become coarser (associated with glaze ice growth). The depth of the fine-grained region is between 2 and 3 mm. The above model prediction for maximum rime ice thickness for these conditions is 2.58 mm.

Water usually first appears at the point where $\dot{m}(x, s, t)$ takes its greatest value. Beyond this time, the model must be treated with care: even if $T < 0$ at a given point, there is now the possibility that water will flow there from a warmer region. The coupling of the flow and thermal problems determines where the glaze or rime ice models apply.

For regions where the surface is covered by water, integration of the pseudo-steady model gives

$$T = \frac{z}{s} - 1 \qquad \theta = \frac{E_{0g}}{1 + E_{1g}h}(z - s)\,. \tag{3.124}$$

The Stefan condition is

$$\frac{\partial s}{\partial t} = \frac{\partial T}{\partial z} - K\frac{\partial \theta}{\partial z} = \frac{1}{s} - K\frac{E_{0g}}{1 + E_{1g}h}\,, \tag{3.125}$$

where $K = k_l/k_s$. Equation (3.125) coupled with the flow equation (3.32) and appropriate boundary conditions determines the unknowns s, h in the glaze region. Outside of this region, the rime model still applies.

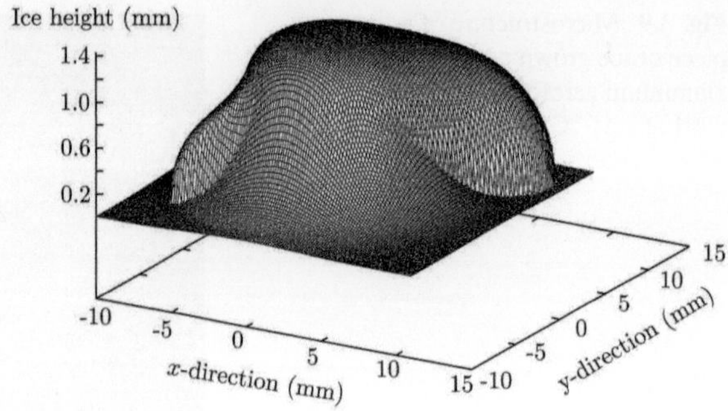

Fig. 3.10 Ice layer, due to an incoming Gaussian distribution of water droplets, on a flat surface. Air shear acts to the right; gravity acts to the left

The three papers [104, 113, 114] deal with flow over a three-dimensional, evolving ice surface. This requires the introduction of a generalised version of Eq. (3.32):

$$\frac{L}{U\tau}\frac{\partial h}{\partial t} + \nabla \cdot \mathbf{Q} = \frac{\dot{m}}{\rho_l} - \frac{\rho_s}{\rho_l}\frac{L}{U\tau}\frac{\partial s}{\partial t}, \tag{3.126}$$

where the fluid flux

$$\mathbf{Q} = \left(-\frac{h^3}{3}\left(\frac{\partial p}{\partial x} - B\hat{\mathbf{g}}\cdot\hat{\mathbf{x}}\right) + \frac{h^2}{2}A_1, -\frac{h^3}{3}\left(\frac{\partial p}{\partial y} - B\hat{\mathbf{g}}\cdot\hat{\mathbf{y}}\right) + \frac{h^2}{2}A_2\right) \tag{3.127}$$

and the pressure depends on the $\hat{\mathbf{z}}$ component of gravity and also surface tension (which is proportional to $\nabla^2 h$). The air and water droplet flow around an aerofoil was modelled using the FLUENT V CFD package in the study of [104].

In Fig. 3.10, an ice shape resulting from the calculations described in [104] is presented. The full calculation conditions and details of the numerical scheme are provided in [104, 113, 114]. The substrate is inclined at an angle 20° to the horizontal, forcing liquid to flow to the left of the figure. This is countered by air shear that acts to the right. Small water droplets impact on the surface $z = 0$ with a Gaussian profile (centred around $(x, y) = (0, 0)$). Initially, rime ice builds up and then a thin liquid layer forms on the top. The ice shape reflects the Gaussian form of the incoming droplets. If all the impinging fluid were to freeze instantaneously when reaching the solid substrate, the ice accretion would have a Gaussian profile with a maximum height $s = 3.5$ mm. The present shape culminates at $s = 1.45$ mm: the difference is due to the water flow.

Figure 3.11 shows the water flow on top of the ice shown in Fig. 3.10. A significant feature of the flow is the capillary ridge on the left-hand side; this is typical of flows where surface tension plays a role. It may be expected that a similar ridge occurs on the right, where air shear has driven the flow. Its absence is due to the ice growth rapidly removing the water there. To actually solve the flow equation (3.126) is a delicate issue, well beyond the scope of this text. Surface

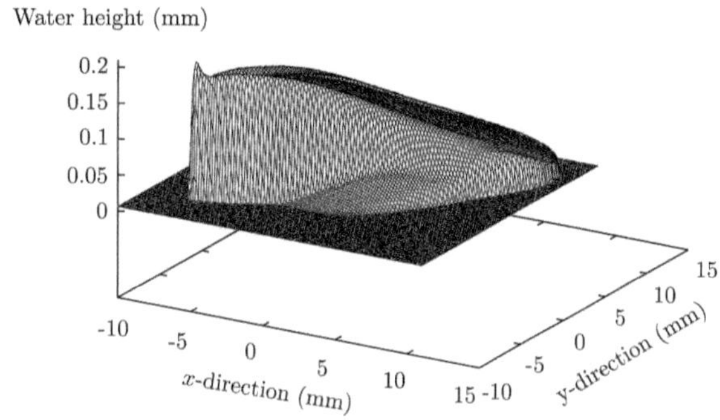

Fig. 3.11 Water layer, due to an incoming Gaussian distribution of water droplets, on a flat surface. Air shear acts to the right; gravity acts to the left

tension-driven thin flow is a complex field; here it is coupled with a moving contact line (an intricate field in itself) as well as including phase change. Properties and applications of thin-film equations are reviewed in [96, 124], and numerical techniques for driven thin-film equations are compared in Ha et al. [60].

Revisiting the base model, for one-dimensional growth, Brakel and Myers [16] extend the analysis to account for a thin substrate (such as the aircraft shell or a thin structure) as well as looking at the perturbation solution for small time derivative terms. This involves solving heat equations in three thin layers. They also note that incoming water droplets may not simply join the liquid layer at the surface but, due to their high velocity, may mix throughout the layer, so adding a source term into the liquid heat equation. For aircraft icing, they note that the high kinetic energy of the droplets increases the liquid temperature significantly, while for structural icing, they make little difference to the energy balance.

3.4.1 Ice Crystal Icing

A modification of the standard aircraft icing problem is studied in [17, 18] to deal with ice crystal icing in aircraft engines. The physical setup involves an incoming ice-water source that lands on a heated engine surface. Initially, all of the ice melts and a thin water layer forms. This layer grows until a time is reached when the free surface achieves the freezing temperature (this may never occur, but then the system is of no concern). Subsequently, a thin ice layer starts to form, but not all of the incoming mass freezes, so it is assumed that the ice is covered by a water layer, i.e. in this second stage, there is a water-ice-water layer system. The problem is tackled following the approach discussed earlier in the aircraft icing section. This is an interesting and important extension to the models discussed in this chapter, since it deals with a different icing mechanism. However, the upper water layer is treated inconsistently. The assumption [18, Assumption A6] states that this layer "is sufficiently thin that wall-normal temperature gradients may be neglected," and consequently, the temperature is assumed to equal the phase change temperature throughout the layer. In contrast in the two lower layers, the thinness is exploited to define the pseudo-steady approximation. A more appropriate justification could be that the incoming mass contains a mixture of ice and water at the melt temperature, and this constitutes the upper water layer.

Peters et al. [126] also study the ice crystal icing problem but now employing the enthalpy method. Under a pseudo-steady assumption, Eq. (2.140) reduces to

$$0 = \frac{\partial}{\partial x}\left(k_l \frac{\partial \theta}{\partial z}\right) . \tag{3.128}$$

Obviously, this is the standard pseudo-steady temperature equation. In the first stage, where there is a pure water layer, the boundary conditions are

$$\theta(0,t) = T_s > 0 , \qquad -k_l \frac{\partial \theta}{\partial z} = E_0 + E_1\theta + (\mathrm{h} - \mathrm{h}_{inc})|_{z=h_T} , \tag{3.129}$$

where h represents the enthalpy; see the definition of Eq. (2.139) (but note there is no pure solid component) and E_0, $E_1\theta$ energy terms similar to those defined in the glaze ice problem, Eq. (3.119). The incoming enthalpy, i.e. that of the incoming ice-water crystals, is a known quantity h_{inc}. The total thickness of the layer comes from a mass balance:

$$\rho \frac{dh_T}{dt} = \dot{m}_{inc} - \dot{m}_{evap} . \tag{3.130}$$

These equations are sufficient to determine the temperature (and hence enthalpy) and film height. For the simple case of approximately constant incoming and evaporating mass rates, $h_T = (\dot{m}_{inc} - \dot{m}_{evap})t/\rho$. This stage lasts until $\theta = T_m$ (at time $t = t_m$). Subsequently, the water layer is covered by a mixed ice-water mush (which therefore has temperature T_m). Now the total height is divided into the original water layer and the mush that consists of ice and water in an as yet unknown proportion.

To determine the water-ice fraction of the mush, we note that since $\theta = T_m$ throughout the mush, the energy flux, $k_l\theta_z$, at the free surface must be zero:

$$0 = E_0 + E_1 T_m + (\mathrm{h} - \mathrm{h}_{inc})|_{z=h_T} \quad \Rightarrow \quad \beta = \frac{\mathrm{h}^*_{inc}}{L_m} = \frac{\mathrm{h}_{inc} - E_0 - E_1 T_m}{L_m} \tag{3.131}$$

This defines the "balancing enthalpy" $h = \mathrm{h}^*_{inc}$, and β is the "mushy phase enthalpy fraction". From this point, the incoming mass is now split by this fraction. For example, with constant incoming and evaporating mass fluxes, the thickness of the mush layer $h_m(t) = (\dot{m}_{inc} - \dot{m}_{evap})(t - t_m)/\rho$ consists of an amount of water equal to the height $h_w = \beta h_m$ and of ice $h_i = (1-\beta)h_m$. For a more detailed explanation, the reader is referred to [126].

3.5 Conclusions

Stefan problems can already be sufficiently difficult to analyse without introducing the added complication of fluid flow. In this chapter, we have attempted to simplify the problem by considering only flows where the fluid layer is thin (where the definition of "thin" is in the sense that lubrication theory holds). This covers a surprisingly large amount of practically interesting situations.

First, two examples of channel flow were dealt with. The first involved solidification in a microchannel, and with careful use of co-ordinate transformations, a separable solution was obtained to describe the temperature field and so reduce the problem to a single integro-differential equation. For the case of contact melting, flow beneath the melting material was driven by the weight of the material. Again separation of variables helped to partially solve the problem, thus demonstrating how basic undergraduate techniques can aid in solving seemingly very difficult problems.

The atmospheric icing study introduced the renowned thin-film equation of fluid mechanics, which has received a wealth of theoretical attention. However, before arriving at the film flow, a model was developed for dry ice growth, which accurately predicts when water first appears. Once water is flowing, this must be solved numerically. Although complex, this is an important physical problem, and the models described in the atmospheric icing section are currently employed in aircraft and structural icing codes around the world.

Finally, the techniques discussed in earlier chapters were shown to apply to a novel form of model, ice crystal icing, leading to a simple and practical solution form. This one-dimensional analysis has the potential to be extended in the same way previous one-dimensional icing models have been extended to include flow and incorporated into icing codes.

Chapter 3 Exercises

1. Starting from the two-dimensional Cartesian form of Eq. (3.1) with constant viscosity, set $\mathbf{u} = (u(z), 0)$ such that we consider steady, unidirectional flow in a channel:

   (a) If the channel lies between $z \in [0, H]$ and no-slip holds at the walls, determine the form of $u(z)$ in terms of the pressure gradient.
   (b) If there is a constant flux Q, determine the pressure along a channel $x \in [0, L]$ (assume the exit pressure is ambient $p|_{x=L} = p_a$).

2. Section 3.1.1 deals with Cartesian channel flow. Follow the same derivation but now for a circular channel.
3. Consider a long line of fluid on a horizontal surface. In the stationary state, Eq. (3.33), with pressure defined by Eq. (3.27), may be used to determine the shape of the free surface by setting $Q = 0$.

   (a) Assuming no surface shear, $A = 0$, (and no solidification $s = 0$), determine the cross-sectional height $z = h(x, t)$ if $h(\pm 1, t) = 0$ and $h_x(1) = -\tan\gamma$

(note γ is known as the contact angle and it depends on the fluid/solid interaction as well as the motion before equilibrium is reached).

(b) Investigate the limits where $B/C \gg 1$ and $B/C \ll 1$, i.e. gravity or surface tension dominates.

4. The early time solution for solidification in a Cartesian channel is given by Eq. (3.61), $s \approx 1-\sqrt{2t}$. Taking this as an approximation valid for all time, plot the evolution of the pressure drop. With a constant, non-zero, flux what happens as $t \to 1/2$ and what is the physical significance?
5. Given that the expression for the solid layer thickness inside a circular pipe is expressed by

$$s\frac{ds}{dt} = \frac{1}{\ln s},$$

determine the form of s. From this calculate the nondimensional closure time.
6. A classic squeeze film consists of a solid slowly displacing a thin fluid layer due to a fixed applied force. This forms part of the contact melting theory. In the absence of a phase change for the two-dimensional case, we have

$$u = \frac{p_x}{2}z(z-h) \qquad w(h,t) = \frac{\partial h}{\partial t}.$$

If the film is being driven by a constant applied force

$$F = \int_{-1}^{1} p\,dx,$$

calculate the evolution of the height $h(t)$ subject to $h(0) = 1$.
7. Starting from the two-dimensional Cartesian version of the heat equations (1.6a, 1.7), derive (3.41). What are the neglected terms?
8. Starting from the three-dimensional heat equations, including flow in the liquid, derive the reduced form (3.63) using the scales defined by (3.74, 3.75).
9. Consider the one-dimensional heat equation $T_t = T_{zz}$ in the context of contact melting. This is subject to $T(h,t) = 0$, $T_z(H,t) = 0$, where $h(t)$ satisfies an appropriate Stefan condition and $H = H(t)$ decreases over time.
Explain why this system cannot be solved using separation of variables.
Note, there frequently appear papers, in reputable journals, where the Stefan problem is incorrectly solved via separation of variables with the above error. Do not fall into this trap!

10. The temperature in a semi-infinite media subject to a Robin condition may be expressed by Eq. (3.102):

$$T = 1 + (T_0 - 1)\left[\operatorname{erf}\left(z\sqrt{\frac{A_1}{4t}}\right) + \exp\left(Bi\, z + t\, Bi^2/A_1\right)\right.$$
$$\left.\times \operatorname{erfc}\left(z\sqrt{\frac{A_1}{4t}} + Bi\sqrt{\frac{t}{A_1}}\right)\right] :$$

    (a) Noting that

    $$\lim_{x\to\infty} \exp(x^2)\operatorname{erfc}(x) = 0,$$

    write down an expression for the temperature $T(z, t)$ in the limit of infinitely high heat transfer. In this case when does melting first occur?
    (b) Determine the TIM solution corresponding to $Bi \to \infty$.
    (c) Choose sensible values of A_1 and plot your two solutions. How do they compare?

11. Equation (3.94) defines

$$1 - h_m(t) = \sum_{n=1}^{\infty} \int_0^1 \int_0^1 \Phi_n(Y, t) \sin(\omega_n X)\, dY\, dX :$$

    (a) Taking Φ_n from (3.93), evaluate the integrals to obtain an equation relating $1 - h_m(t)$ to $f(t)$ and the eigenvalues.
    (b) Now take just the first term of the series to determine Λ in Eq. (3.95).
    (c) The two-dimensional contact melting problem is retrieved by setting $\lambda = 0$. From the solution to part (b), derive the two equation model equivalent to (3.115).

12. For the aircraft icing problem, for sufficiently small times such that $h, s \ll 1$, the evolution of the layers is defined by

$$\frac{\partial s}{\partial t} = \frac{1}{s} - A \qquad \frac{\partial h}{\partial t} = \frac{U\tau}{\rho_l L}\dot{m} - \frac{\rho_s}{\rho_l}\frac{\partial s}{\partial t}.$$

    Calculate s, h when $\dot{m}$ is constant (assume $h = s = 0$ when $t = 0$).
13. Following the approach of [126], determine an expression for the time when mush first starts to appear if the incoming and evaporating mass rates are constant.

Chapter 4
Variable Thermophysical Properties and Phase Change Temperature

4.1 Introduction

The governing equations discussed in Chap. 1 were derived for highly idealised situations. They constitute the classical form of Stefan problem and, in combination with the boundary conditions of Sect. 1.2.1, have been applied to describe a wide variety of physical phenomena. However, in practice, they may be too idealised, exhibiting unrealistic features and incapable of capturing observed physical effects. For example, a fixed boundary temperature, different to the initial value, represents an instantaneous switch (and we have already discussed the infinite velocity resulting from this condition). In practice, this is impossible to achieve; even away from the initial switch, it is difficult to maintain a constant boundary temperature throughout a phase change process. The equations of Sect. 1.2 are based on the assumption of constant thermophysical properties, for example, we write the thermal diffusion term as kT_{xx}, but with a variable conductivity, this should be $(kT_x)_x$. In practice, properties are temperature dependent (admittedly sometimes only weakly so), and they invariably differ between phases (often quite significantly). There also exist situations where properties are size or velocity dependent. This may be observed, for example, when the curvature is high, such as in dendritic growth, during nanoscale phase change, or with the solidification of supercooled materials [4, 40, 52, 116].

In this chapter, we will generalise the governing equations prescribed in earlier chapters to account for a number of nonclassical effects. The derivation will start from standard conservation equations, as detailed in Bird et al. [15], and follow the route of Myers et al. [119].

T. G. Myers, *Theoretical and Practical Stefan Problems*, Interdisciplinary Applied Mathematics 63, https://doi.org/10.1007/978-3-032-04826-4_4

4.2 Governing Equations for Phase Change

Phase change involves the energy-driven transformation of mass from one form into another. It is therefore natural to formulate the governing equations in terms of conservation laws and the thermodynamics of a two-phase mixture. The appropriate conservation equations are provided in [15]. In the specific context of phase change, they are applied in [4, 102, 119].

Here we start with the following conservation laws:

$$\frac{\partial \rho}{\partial t} + \nabla \cdot (\rho \mathbf{v}) = 0\,, \tag{4.1a}$$

$$\frac{\partial}{\partial t}(\rho \mathbf{v}) + \nabla \cdot \left(\rho \mathbf{v} \otimes \mathbf{v} + p\underline{\mathbf{I}}\right) = 0\,, \tag{4.1b}$$

$$\frac{\partial}{\partial t}\left(\frac{1}{2}\rho v^2\right) + \nabla \cdot \left(\frac{1}{2}\rho v^2 \mathbf{v} + p\mathbf{v}\right) - p\nabla \cdot \mathbf{v} = 0\,, \tag{4.1c}$$

$$\frac{\partial}{\partial t}\left(\rho\left[u + \frac{1}{2}v^2\right]\right) + \nabla \cdot \left(\rho\left[u + \frac{1}{2}v^2\right]\mathbf{v} + \mathbf{q} + p\mathbf{v}\right) = 0\,, \tag{4.1d}$$

which represent conservation of mass, momentum, mechanical energy, and total energy; see [15, §3.1, 3.2, 3.3, 10.1], [4, §2.3E]. The various parameters are density ρ, velocity $\mathbf{v}$ (where $v = |\mathbf{v}|$), pressure p, internal energy per unit mass u, and the conductive heat flux $\mathbf{q}$. Equation (4.1a) is a mass balance. Equation (4.1b) states that the change in momentum is balanced by the movement of momentum and the effect of pressure, where $\otimes$ is the tensor product and $\underline{\mathbf{I}}$ the unit tensor. Since we will be working in three-dimensional space, we may define $\mathbf{v} \otimes \mathbf{v} = \mathbf{v}\mathbf{v}^T$, which is a symmetric 3×3 matrix. Equation (4.1b) therefore differs from the rest in that it is a vector conservation law rather than a scalar form. Equation (4.1c) states that the change in kinetic energy is balanced by the input of kinetic energy through the bulk flow; the work done by pressure, p; and the rate of reversible conversion to internal energy. Equation (4.1d) states that the rate of change of total energy depends on energy flow through movement, conduction, and the work done by pressure.

In writing down these equations, we have assumed that gravity and viscous effects are negligible. To simplify the discussion in the following, we will work in the context of a solid to liquid transition but the model will, of course, be valid for other transitions. We assume that the solid and liquid phases are incompressible, having constant densities ρ_s and ρ_l that, in general, will not be equal $\rho_s \neq \rho_l$. For the equations to hold throughout the material requires an assumption that there exists at least a small region, which defines the interface, where there is a smooth transition from solid to liquid (equivalent to the mushy region discussed in Sect. 1.2.2). In the following, we impose a sharp-interface limit, that is, the width of the interface tends to zero.

In regions of constant density, i.e. the solid and liquid zones, Eq. (4.1a) reduces to

$$\nabla \cdot \mathbf{v} = 0 . \tag{4.2}$$

Consequently, the final term of Eq. (4.1c) is zero within each region. Subtracting equation (4.1c) from Eq. (4.1d) leads to the equation for conservation of internal energy:

$$\rho \left(\frac{\partial u}{\partial t} + \mathbf{v} \cdot \nabla u \right) = -\nabla \cdot \mathbf{q} . \tag{4.3}$$

The standard heat equation is often derived from this balance by expressing u and $\mathbf{q}$ in terms of the temperature. These are discussed later.

4.2.1 Jump Conditions at the Interface

The density and internal energy are discontinuous across a sharp solid-liquid interface. Conservation of mass and energy links the jump in these two quantities to the motion of the interface.

For a scalar conservation law written in divergence form,

$$\frac{\partial F}{\partial t} + \nabla \cdot \mathbf{G} = 0, \tag{4.4}$$

the Rankine-Hugoniot condition states that the jump in F across a discontinuity moving with normal velocity v_n is given by

$$[F]_-^+ v_n = [\mathbf{G} \cdot \hat{\mathbf{n}}]_-^+, \tag{4.5}$$

where $\hat{\mathbf{n}}$ is the unit vector normal to the discontinuity, v_n represents the normal velocity of the interface, and $[\cdot]_-^+$ denotes the jump. In one-dimensional Cartesian problems, the interface is generally denoted by $s(t)$ and $v_n = ds/dt$ and in one-dimensional cylindrical and radial problems by $R(t)$ and $v_n = dR/dt$. With more dimensions, more care is required, for example, in a two-dimensional Cartesian system, if the interface is defined by $z = s(x, y, t)$, then

$$v_n = (0, 0, s_t) \cdot \frac{(-s_x, -s_y, 1)}{\sqrt{1 + s_x^2 + s_y^2}} = \frac{s_t}{\sqrt{1 + s_x^2 + s_y^2}} . \tag{4.6}$$

The derivation of Eq. (4.5) follows simply from integrating equation (4.4) across a finite region, which includes the jump and then letting the thickness of this region tend to zero. In the following examples, the normal vector points into the liquid so that the jump is measured from the solid into the liquid, i.e. $[\cdot]_-^+ = [\cdot]_s^l$.

The Rankine-Hugoniot condition, (4.5), may be applied immediately to the mass and energy balances (4.1a, 4.1d) to give

$$(\rho_l - \rho_s)v_n = (\rho_l \mathbf{v}_l - \rho_s \mathbf{v}_s) \cdot \hat{\mathbf{n}}, \quad (4.7a)$$

$$\rho_l \left(u_l + \frac{v_l^2}{2} \right) (v_n - \mathbf{v}_l \cdot \hat{\mathbf{n}}) - \rho_s \left(u_s + \frac{v_s^2}{2} \right) (v_n - \mathbf{v}_s \cdot \hat{\mathbf{n}}) = $$
$$[(\mathbf{q}_l - \mathbf{q}_s) + (p_l \mathbf{v}_l - p_s \mathbf{v}_s)] \cdot \hat{\mathbf{n}}, \quad (4.7b)$$

where all variables are evaluated at the solid-liquid interface. We previously encountered a form of (4.7a) in Sect. 3.1 where it was instead obtained through a mass balance, Eq. (3.18). The equivalence to Eq. (3.18) is apparent after noting that in two dimensions, the normal velocity $v_n = \mathbf{u_b} \cdot \hat{\mathbf{n}} = s_t/\sqrt{1+s_x^2}$. In Sect. 3.1, the denominator was neglected consistent with the assumption that gradients are small, $s_x^2 \ll 1$; however, in general, it may require inclusion. Equation (4.7b) is an extension of the classical Stefan condition, Eq. (1.2), accounting for kinetic energy and pressure-volume work. The one-dimensional version of this Stefan condition is derived through conservation laws in [4, §2.3E], [106].

We can also apply the Rankine-Hugoniot condition to each component of the momentum balance (4.1b) and then combine the results to obtain a vector jump condition:

$$\begin{aligned}(\rho_l \mathbf{v}_l - \rho_s \mathbf{v}_n)\, v_n &= \left(\rho_l \mathbf{v}_l \otimes \mathbf{v}_l + p_l \underline{\mathbf{I}}\right) \cdot \hat{\mathbf{n}} - \left(\rho_s \mathbf{v}_s \otimes \mathbf{v}_s + p_s \underline{\mathbf{I}}\right) \cdot \hat{\mathbf{n}} \\ &= \rho_l (\mathbf{v}_l \cdot \hat{\mathbf{n}})\mathbf{v}_l - \rho_s (\mathbf{v}_s \cdot \hat{\mathbf{n}})\mathbf{v}_s + (p_l - p_s)\hat{\mathbf{n}}. \end{aligned} \quad (4.8)$$

The second relation is obtained after applying $\underline{\mathbf{I}} \cdot \hat{\mathbf{n}} = \hat{\mathbf{n}}$ and $(\mathbf{v} \otimes \mathbf{v}) \cdot \hat{\mathbf{n}} = (\mathbf{v} \cdot \hat{\mathbf{n}})\mathbf{v}$. Normal and tangential jump conditions can then be obtained by taking the appropriate dot products of (4.8) to give

$$\left[(\rho_l \mathbf{v}_l - \rho_s \mathbf{v}_s) \cdot \hat{\mathbf{n}}\right] v_n = \rho_l (\mathbf{v}_l \cdot \hat{\mathbf{n}})^2 - \rho_s (\mathbf{v}_s \cdot \hat{\mathbf{n}})^2 + p_l - p_s, \quad (4.9a)$$

$$\left[(\rho_l \mathbf{v}_l - \rho_s \mathbf{v}_s) \cdot \hat{\mathbf{t}}\right] v_n = \rho_l (\mathbf{v}_l \cdot \hat{\mathbf{n}})(\mathbf{v}_l \cdot \hat{\mathbf{t}}) - \rho_s (\mathbf{v}_s \cdot \hat{\mathbf{n}})(\mathbf{v}_s \cdot \hat{\mathbf{t}}), \quad (4.9b)$$

where $\hat{\mathbf{t}}$ is the unit tangent vector.

In [4], a number of cases of solidification and melting of a semi-infinite one-dimensional material are described. The interface velocity is written as $v_n = s_t$ and $\mathbf{v}_l \cdot \hat{\mathbf{n}} = v_l$ is the liquid velocity in the x-direction. Here we consider an example where the solid is fixed such that $\mathbf{v}_s = \mathbf{0}$ (and if the solid is not fixed, we can easily switch to a co-ordinate system moving with the solid to obtain an equivalent system). The one-dimensional solid and liquid heat fluxes are given by $q_s = \mathbf{q}_s \cdot \hat{\mathbf{n}}$ and $q_l = \mathbf{q}_l \cdot \hat{\mathbf{n}}$, respectively. The jump conditions (4.7a, 4.7b, 4.9a) for mass, energy, and momentum may then be written as

$$(\rho_l - \rho_s)s_t = \rho_l v_l, \tag{4.10a}$$

$$\rho_l \left(u_l + \frac{v_l^2}{2}\right)(s_t - v_l) - \rho_s u_s s_t = q_l - q_s + p_l v_l\,, \tag{4.10b}$$

$$p_s - p_l = \rho_l v_l (v_l - s_t) = -\rho_s v_l s_t\,, \tag{4.10c}$$

and they hold at $x = s(t)$. In [4, §2.3, eqs 25-27], mass, momentum, and energy jump conditions are written down; the above equations correspond to these in the case where the solid defines the frame of reference such that $v_s = 0$.

The momentum condition (4.10c) highlights a problem with this derivation form. It demonstrates that if the density varies between the phases, then the resultant change in velocity causes a change in the pressure. Physically, this makes sense; we may think of the pressure jump as the equal and opposite force (per unit area) to the kick required to accelerate the fluid. This equation is fine when dealing with a flat interface but something is lacking in the case of a curved interface. For curved interfaces, it is well known that there is also a pressure jump caused by the surface tension $p_s - p_l = \sigma_{sl}\kappa$, where κ is the curvature and σ_{sl} the surface tension between the solid and liquid phases: Eq. (4.10c) misses this contribution. The problem can be traced to the requirement that the conservation equations hold within each domain but do not account for pure boundary effects such as surface tension.

To correctly account for surface tension, a Korteweg stress tensor could be introduced into the conservation equations. This involves a term related to surface energy; it is chosen such that in the limit of a sharp interface, the correct surface energy term appears. This approach is discussed in [119]; however, due to the difficulty in correctly choosing the stress tensor, they finally opt for the simpler "fix" of adding a source term into the conservation equations, which accounts for surface tension:

$$\frac{\partial}{\partial t}(\rho \mathbf{v}) + \nabla \cdot \left(\rho \mathbf{v} \otimes \mathbf{v} + p\underline{\mathbf{I}}\right) + \sigma\kappa\delta(x - s)\hat{\mathbf{n}} = 0\,, \tag{4.11a}$$

$$\frac{\partial}{\partial t}\left(\frac{1}{2}\rho v^2\right) + \nabla \cdot \left(\frac{1}{2}\rho v^2 \mathbf{v} + p\mathbf{v}\right) - p\nabla \cdot \mathbf{v} + \sigma\kappa\delta(x - s)v_n = 0\,, \tag{4.11b}$$

$$\frac{\partial}{\partial t}\left(\rho\left[u + \frac{1}{2}v^2\right]\right) + \nabla \cdot \left(\rho\left[u + \frac{1}{2}v^2\right]\mathbf{v} + \mathbf{q} + p\mathbf{v}\right) + \sigma\kappa\delta(x - s)v_n = 0\,, \tag{4.11c}$$

where δ is the Dirac delta function. The δ function ensures that surface tension only contributes at the interface. The extension to multiple interfaces is easily achieved with this approach by simply adding more source terms.

As stated earlier, the Rankine-Hugoniot condition may be derived by integrating the governing equations across a finite region, which includes the jump and then letting the thickness of this region tend to zero. Noting that $\int_{-a}^{a} \delta(x) f(x)\, \mathrm{d}x = f(0)$ for any $a > 0$ and that the velocity v_n may be written s_t, the momentum and total energy equations (4.10c, 4.10b) may now be written as

$$p_s - p_l = \sigma_{sl}\kappa - \rho_s v_l s_t, \tag{4.12a}$$

$$\rho_l \left(u_l + \frac{v_l^2}{2} \right) (s_t - v_l) - \rho_s u_s s_t = q_l - q_s + p_l v_l + \sigma_{sl}\kappa s_t\,, \tag{4.12b}$$

where all quantities are defined at $x = s(t)$. In the case of a flat interface, $\kappa = 0$, Eqs. (4.10c, 4.10b) are retrieved. The pressure jump now correctly accounts for surface tension, but we note that surface tension also appears in the energy balance (it has no effect on the mass balance (4.10a)). Equation (4.12b) may be rearranged, after invoking equation (4.10a), to give a form of Stefan condition for a stationary solid phase process that accounts for density difference and curvature:

$$\left[\rho_s \left(u_l - u_s + \frac{v_l^2}{2} \right) - \sigma_{sl}\kappa + \left(\frac{\rho_s}{\rho_l} - 1 \right) p_l \right] \frac{\partial s}{\partial t} = q_l - q_s\,; \tag{4.13}$$

again all quantities are defined at $x = s$. This may be easily modified for different liquid and solid velocity combinations.

4.2.2 *Internal Energy and Heat Flux*

The Stefan problem is most commonly written in terms of temperature rather than internal energy and heat flux. The internal energy per unit mass (or specific internal energy) depends on the heat energy and work done against an external pressure. The heat energy is related to temperature by the heat capacity, c, which is defined as the ratio of heat energy entering a system to the temperature change induced by that energy $c = q/\Delta T$ (if c varies with temperature, then we define $q = \int c\, \mathrm{d}T$). It is standard to discuss heat capacities measured either at constant pressure, c_p, or constant volume, c_v. For solids and liquids, which are virtually incompressible, $c_p \simeq c_v$, so in the following, we make no distinction between the two and simply write c. To convert from internal energy to temperature, we follow [4, 15] and write

$$u_s = c_s(T - T_m) - \frac{p_s}{\rho_s}\,, \tag{4.14a}$$

$$u_l = c_l(\theta - T_m) + L_m - \frac{p_l}{\rho_l}\,. \tag{4.14b}$$

This definition is consistent with the enthalpy and internal energy terms defined in the discussion of mushy regions in Sect. 1.2.2. The bulk phase change temperature T_m is measured at a specified reference pressure, denoted p_{ref}. This temperature may be thought of simplistically as the one we are most familiar with, for example, with water $T_m = 273.15$ K, and this occurs at the reference pressure of 1 atmosphere or 101.325 kPa. It is an arbitrary but standard choice to use the bulk value in the conversion from energy to temperature. This choice means that in most Stefan problems, when the temperature reaches the bulk phase change value (for the specified pressure), the enthalpy, $h_s = c_s(T - T_m)$, of the solid is zero, and the jump between the liquid and solid enthalpies is the latent heat. In [4], a reference internal energy is added to the problem $u_{\text{ref}} = p_{\text{ref}}/\rho_s$ such that $u(p_{\text{ref}}, T_m) = 0$. However, given that it is a reference state, it cancels throughout the calculations; hence, for simplicity, we will omit this term.

In [15], a differential form for the internal energy is provided, $\mathrm{d}u = c_v \mathrm{d}T$. For incompressible materials, the volume is constant in each phase and we may employ this relation in deriving the heat equation. Note, in [4], a slightly different route is followed to obtain $\mathrm{d}u = c_p \mathrm{d}T$ (recall that subscripts v and p denote fixed volume and fixed pressure) despite the fact that pressure may vary in the phases. However, since the phases are incompressible and so the specific heats are equivalent, the result is equivalent, and we may write

$$\mathrm{d}u = c\,\mathrm{d}T\,. \tag{4.15}$$

The standard form of heat flux is based on Fourier's law:

$$\mathbf{q} = -k\nabla T\,, \tag{4.16}$$

where k is the thermal conductivity. The heat equations may be obtained from the conservation of internal energy, Eq. (4.3), by noting that $\mathrm{d}u = c\,\mathrm{d}T$ and using the appropriate law for the heat flux; in this case, (4.16). Since the material properties are constant within each phase, we obtain

$$\rho_s c_s \frac{\partial T}{\partial t} = k_s \nabla^2 T, \qquad 0 < x < s\,, \tag{4.17}$$

$$\rho_l c_l \left(\frac{\partial \theta}{\partial t} + \mathbf{v}_\mathbf{l} \cdot \nabla\theta\right) = k_l \nabla^2 \theta, \qquad s < x\,. \tag{4.18}$$

These equations may also be obtained by substituting for u, via Eqs. (4.14a, 4.14b), into Eq. (4.3). This leads to the inclusion of terms of the form $\mathrm{d}(p/\rho)/\mathrm{d}t$. In [15, §11.2], these are shown to be zero for incompressible phases.

Using definitions (4.14a, 4.14b) and denoting the interface temperature $T(s(t), t) = \theta(s(t), t) = T_I(t)$ (where T_I does not necessarily equal to T_m), the Stefan condition (4.13) may now be written as

$$\left[\rho_s L_m + \rho_s(c_l - c_s)(T_I - T_m) + p_s - p_l - \sigma_{sl}\kappa + \frac{\rho_s v_l^2}{2}\right]\frac{\partial s}{\partial t} = k_s \nabla T \cdot \hat{\mathbf{n}} - k_l \nabla\theta \cdot \hat{\mathbf{n}}\,, \tag{4.19}$$

Equation (4.12a) permits the pressure and surface tension terms to be replaced by $-\rho_s v_l s_t$ and finally (4.10a) expresses v_l in terms of s_t to arrive at the new form of Stefan condition:

$$\rho_s\left[L_m + (c_l - c_s)(T_I - T_m) + \frac{1}{2}\left(\left(\frac{\rho_s}{\rho_l}\right)^2 - 1\right)\left(\frac{\partial s}{\partial t}\right)^2\right]\frac{\partial s}{\partial t} = k_s \nabla T \cdot \hat{\mathbf{n}} - k_l \nabla\theta \cdot \hat{\mathbf{n}}\,. \tag{4.20}$$

Equations (4.17, 4.18, 4.20) define the two phase Stefan problem when there is a density difference between the phases, a stationary solid and an interface temperature (which may be time dependent) not equal to the bulk value. Kinetic energy may be observed to affect the energy available for phase change, and this depends on the size of ρ_s/ρ_l – a positive requirement of kinetic energy will slow down the phase change; however, if the contribution is negative, then the process may be faster (i.e. motion induced by a change in density could add energy to the system). Recall the Neumann solution presented following Eq. (2.15) accounted only for liquid motion but not the complication of the associated energy change.

Interestingly, the surface tension (or surface energy) contribution, which is apparent in Eq. (4.13), no longer appears in the final Stefan condition. It has cancelled out through the contribution of pressure to the internal energy. In fact, it has not completely disappeared; it can influence the phase change temperature T_I, specifically when the Gibbs-Thomson relation is invoked, which relates the interface temperature to the curvature. This is discussed later in Sect. 5.1.

Neglect of kinetic energy from the Stefan condition is common, either because it is small compared to the latent heat or because density variation is not considered. However, it is worth pointing out that when applying a fixed-temperature boundary condition, $T(0,t) = T_0$, then the initial solidification velocity is infinite (recall the Neumann solution of Sect. 2.1.1 where $s_t = \lambda t^{-1/2}$) and kinetic energy dominates. Also in the final stages of melting, a cylindrical or spherical material from the outside the melt rate tends to infinity. Hence, there are times when kinetic energy is dominant, yet it is omitted from virtually all analyses; see the discussion in [53].

It is important to note that Eq. (4.20) does not exactly coincide with related Stefan conditions in previous textbooks.

Davis [40, Eq. 2.62] presents a reduced version of Eq. (4.19) for a growing curved interface with no change in density or specific heat, resulting in a coefficient of s_t in the Stefan condition involving only the latent heat and surface tension – Eq. (4.19) cannot be reduced to this form. By choosing $\rho_s = \rho_l$ (and hence $v_l = 0$), $c_s = c_l$, the kinetic energy and sensible heat terms drop out of (4.19). However, in [40], the pressure-work term is omitted in the energy balance (i.e. the $p_s - p_l$ terms do not appear). If pressure-work were to be included, then it would cancel with surface tension, and a standard Stefan condition, valid for fixed density and specific heat, is retrieved.

Equation (4.20) also does not match the form provided in the textbook of Alexiades and Solomon [4, §2.3]. We mention this here since their equations have been taken up by subsequent authors, unfortunately including my group; see, for example [53, 132, 137]. The issue may be traced to the solution of their Problem 24, §2.3. Their equation (48a, §2.3) defines the internal energy in the solid at the interface as the reference value; this requires choosing the solid pressure $p_s(s) = p_{\text{ref}}$. To derive their equation (48b), they neglect a term in the interface value of the internal energy of the liquid; to do this requires $p_l(s) = p_{\text{ref}}$, i.e. $p_l(s) = p_s(s)$. Yet their equation (33) defines the pressure jump $p_l(s) - p_s(s) \neq 0$. The goal of their equation (37) is to determine $p_s(s) - p_{\text{ref}}$, which then introduces an erroneous kinetic energy term, so affecting the expression in the final equation and thus all expressions of unfortunate subsequent authors who took this as their base model. However, bear in mind that this is a complex problem and these books are generally excellent!

4.3 Application to Standard Configurations

4.3.1 One-Dimensional Solidification of a Semi-infinite Fluid

To illustrate how to use the equations derived in the previous section, we begin with the standard problem of the solidification of a semi-infinite one-dimensional fluid driven by cooling at $x = 0$. The solid region occupies $x \in [0, s(t))$; the fluid occupies $x > s(t)$, where $s(0) = 0$. In this scenario, it makes sense to keep the solid material fixed and allow the fluid to move due to the density change, so that $\mathbf{v}_s = \mathbf{0}$, and we may apply the equations without any amendment for a moving solid. The interface velocity is written $v_n = s_t$.

Firstly, in this one-dimensional problem, we see that Eqs. (4.17, 4.18) reduce to

$$\rho_s c_s \frac{\partial T}{\partial t} = k_s \frac{\partial^2 T}{\partial x^2}, \qquad 0 < x < s(t)\,, \tag{4.21}$$

$$\rho_l c_l \left(\frac{\partial \theta}{\partial t} + v_l \frac{\partial \theta}{\partial x} \right) = k_l \frac{\partial^2 \theta}{\partial x^2}, \qquad s(t) < x\,. \tag{4.22}$$

The liquid energy equation (4.18) requires knowledge of the velocity $v_l(x, t) = \mathbf{v}_l \cdot \hat{\mathbf{n}}$, where $\hat{\mathbf{n}} = \hat{\mathbf{x}}$. This arises through mass conservation equation (4.2), which here reduces to $\partial v_l / \partial x = 0$ and hence $v_l(x, t) = v_l(t)$. Since the liquid velocity is independent of x, we only require its value at a single position to determine the value throughout the liquid. It may be calculated from the jump condition at $x = s(t)$, Eq. (4.10a):

$$v_l(t) = \left(1 - \frac{\rho_s}{\rho_l}\right)\frac{ds}{dt}\,. \tag{4.23}$$

This in turn depends on the interface velocity, s_t, which is determined by the Stefan condition (4.20):

$$\rho_s\left[L_m + (c_l - c_s)(T_I - T_m) + \frac{1}{2}\left(\left(\frac{\rho_s}{\rho_l}\right)^2 - 1\right)\left(\frac{ds}{dt}\right)^2\right]\frac{ds}{dt} = k_s\frac{\partial T}{\partial x} - k_l\frac{\partial \theta}{\partial x}\,. \tag{4.24}$$

The system of Eqs. (4.21, 4.22, 4.24) should be contrasted with the basic Stefan model of (1.1a–1.2). The Stefan condition, in particular, is significantly more complex and is now cubic in s_t. Of course, we hope that the basic model is a limit of this more complex one. It may be retrieved in the (artificial) limit where $\rho_s = \rho_l$, such that the phase change causes no motion, $v_l = 0$, which then removes the added kinetic energy term. We also need the phase change to occur at the bulk value $T_I = T_m$. So far, we have only studied cases where the interface temperature is fixed at the bulk value, but there exist a number of physically important situations where $T_I \neq T_m$. In Chap. 5, we examine two types of variable interface problem involving supercooling and phase change with high curvature.

4.3.2 *Spherically Symmetric Melting*

We now examine the case of spherically symmetric melting; the physical situation is depicted in Fig. 4.1.

The material is envisioned as being surrounded by a hot ambient environment with temperature T_H. As the solid melts, two interfaces evolve; these are the solid-liquid and the liquid-ambient. Due to the density difference between the solid and liquid, the liquid-ambient interface will grow in order to accommodate the change in volume that occurs during melting. The solid and liquid occupy the domains given by $0 < r < R(t)$ and $R(t) < r < R_b(t)$. In this geometry, the velocity of the solid-liquid interface is denoted $v_n = \mathrm{d}R/\mathrm{d}t$, and the unit normal vector to the interface is $\hat{\mathbf{n}} = \hat{\mathbf{r}}$.

The heat equations are defined by the spherically symmetric versions of (4.17)–(4.18). As before, an expression for the liquid velocity is obtained first by integrating the mass conservation equation (4.2):

$$\frac{1}{r^2}\frac{\partial}{\partial r}\left(r^2 v_l\right) = 0 \quad \Rightarrow \quad v_l(r,t) = A(t)/r^2\,. \tag{4.25}$$

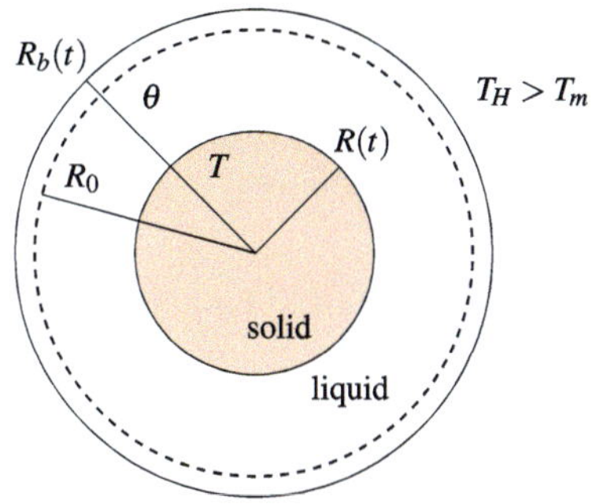

Fig. 4.1 Spherically symmetric melting configuration. The regions of solid and liquid phases are defined by $0 < r < R(t)$ and $R(t) < r < R_b(t)$, respectively, where r is the radial coordinate and t is the time. The melting process is driven by an influx of thermal energy from the surrounding environment, at $T = T_H$, into the liquid phase

The jump condition (4.7a) at $r = R(t)$ gives

$$(\rho_l - \rho_s)\frac{dR}{dt} = \rho_l v_l(R, t) \,. \tag{4.26}$$

Substituting for $v_l(R, t) = A(t)/R^2$ into (4.26) determines the form of $A(t)$, and so we find the relation between the liquid velocity and the rate at which the front moves:

$$v_l(r, t) = \left(1 - \frac{\rho_s}{\rho_l}\right)\frac{R^2}{r^2}\frac{dR}{dt} \,. \tag{4.27}$$

This may be used to specify the velocity in the liquid heat Eq. (4.18).

At the outer boundary, the liquid velocity is simply $v_l(R_b, t) = \mathrm{d}R_b/\mathrm{d}t$. Noting that if we set $r = R_b$, $v_l(R_b, t) = \mathrm{d}R_b/\mathrm{d}t$ in (4.27) and integrate, then the following relation is obtained:

$$R_b^3 = \frac{\rho_s}{\rho_l}R_0^3 + \left(1 - \frac{\rho_s}{\rho_l}\right)R^3 \,, \tag{4.28}$$

where we have imposed $R_b(0) = R(0) = R_0$. This determines the position of the outer boundary in terms of $R(t)$.

The interface velocity is defined by the spherically symmetric version of the Stefan condition (4.20):

$$\rho_s\left[L_m + (c_l - c_s)(T_I - T_m) + \frac{1}{2}\left(\left(\frac{\rho_s}{\rho_l}\right)^2 - 1\right)\left(\frac{dR}{dt}\right)^2\right]\frac{dR}{dt}$$
$$= k_s\frac{\partial T}{\partial r} - k_l\frac{\partial \theta}{\partial r} \,. \tag{4.29}$$

As always, the model is completed by appropriate boundary and initial conditions as well as a form for T_I. In the following section, we will nondimensionalise this system. To do this requires knowledge of the energy driving the phase change, so we now choose to impose a cooling condition:

$$-k\frac{\partial \theta}{\partial r} = \mathscr{H}_{la}(T_H - \theta)\,, \tag{4.30}$$

at $r = R_b(t)$, where $\mathscr{H}_{la}$ is the liquid-ambient heat transfer coefficient. This is augmented with a boundedness condition at the origin:

$$\left.\frac{\partial T}{\partial r}\right|_{r=0} = 0\,, \tag{4.31}$$

and continuity of temperature at the phase change front:

$$T(R,t) = \theta(R,t) = T_I(R)\,. \tag{4.32}$$

The initial condition is simply the initial temperature of the sphere. A common simplification is to set this to the phase change temperature $T(r,0) = T_I$. In this way, we are free to focus purely on the phase change process. Otherwise, we would first have to solve the initial heating phase to determine when $T(R_b,t) = T_I$ – this involves standard heat flow in a fixed domain and so gives us no information with regard to the phase change details.

4.4 Nondimensionalisation

The switch from the standard Stefan condition to the new form of (4.29) is quite dramatic. The relative importance of the new terms in the Stefan problem can be best assessed by casting the model into dimensionless form. As an example, we will now examine the spherical case and define

$$t = \tau\hat{t}\,, \quad r = R_0\hat{r}\,, \quad \hat{T} = \frac{T - T_m}{\Delta T}\,, \quad \hat{\theta} = \frac{\theta - T_m}{\Delta T} \tag{4.33}$$

where the time, length, and temperature scales are τ, R_0, ΔT respectively. These choices are not unique and should be based on the physical situation: here the process is driven by the heat input at the liquid boundary, and so we choose a time-scale based on heat flow through the liquid, $\tau = R_0^2/\alpha_l$, and the temperature scale $\Delta T = T_H - T_m$. The length-scale is the initial sphere radius, R_0.

Immediately dropping the hat notation, the heat equations become

$$\frac{\partial T}{\partial t} = \alpha \frac{1}{r^2} \frac{\partial}{\partial r} \left(r^2 \frac{\partial T}{\partial r} \right), \qquad 0 < r < R(t) , \tag{4.34}$$

$$\frac{\partial \theta}{\partial t} + v_l \frac{\partial \theta}{\partial r} = \frac{1}{r^2} \frac{\partial}{\partial r} \left(r^2 \frac{\partial \theta}{\partial r} \right), \qquad R(t) < r < R_b(t) , \tag{4.35}$$

where $\alpha = \alpha_s/\alpha_l$ and the velocity has been scaled with $V = R_0/\tau$. The Stefan condition becomes

$$\rho \left[\beta + (1-c)T_I + \delta \left(\frac{dR}{dt} \right)^2 \right] \frac{dR}{dt} = \left(k\frac{\partial T}{\partial r} - \frac{\partial \theta}{\partial r} \right) \Bigg|_{r=s(t)} , \tag{4.36}$$

where

$$\rho = \frac{\rho_s}{\rho_l}, \quad \beta = \frac{L_m}{c_l \Delta T}, \quad c = \frac{c_s}{c_l}, \quad \delta = \frac{\rho_l \left(\rho^2 - 1\right) \alpha_l^3}{2R_0^2 k_l \Delta T}, \quad k = \frac{k_s}{k_l}. \tag{4.37}$$

Note T_I is now nondimensional – if we revert to the hat notation to distinguish nondimensional quantities $\hat{T}_I = (T_I - T_m)/\Delta T$, where in the presence of supercooling the interface temperature $\hat{T}_I < 0$, for superheating $\hat{T}_I > 0$.

The dimensionless position of the outer boundary given by Eq. (4.28) is now

$$R_b = \left(\rho - R^3(\rho - 1) \right)^{1/3} , \tag{4.38}$$

and the velocity of the liquid, from Eq. (4.27), is

$$v_l(r,t) = (1-\rho) \frac{R^2}{r^2} \frac{dR}{dt}. \tag{4.39}$$

The boundedness condition at the origin and continuity of temperature remain unchanged. The Newton cooling condition may be written as

$$\frac{\partial \theta}{\partial r} \bigg|_{r=R_b(t)} = Nu[1 - \theta(R_b(t), t)], \tag{4.40}$$

where $Nu = \mathscr{H}_{la} R_0/k_l$ is the Nusselt number.

4.5 Energy Conservation

As discussed in Sect. 2.1.3, the one-phase Stefan problem is a common simplification, usually invoked for the sake of mathematical simplicity. When the melt temperature is the standard phase change value, then the one-phase problem is well defined. However, when the phase change temperature is variable, then difficulties may arise. For example, in the one-phase problem studied in Sect. 5.2 where the liquid is supercooled, it will be shown that by fixing the interface temperature $T_I = 0$, the interface speed tends to infinity as the Stefan number $\beta \to 1^+$, while no solutions exist for $\beta < 1$ [40]. A second issue is that if one phase is neglected, then energy may not be conserved in the supercooled system.

In this section, we will demonstrate that problems with the one-phase reduction may arise due to inconsistent assumptions concerning the temperature in the neglected phase. If the reduction is carried out consistently, then there is no problem with the energy conservation. To keep the analysis simple, we will focus on the situation where the density, ρ, is constant throughout the process. However, the arguments may be easily adapted to include it using the equations described in [53].

We begin by writing down the Cartesian version of Eqs. (4.34, 4.36) in the absence of fluid motion:

$$\frac{\partial T}{\partial t} = \alpha \frac{\partial^2 T}{\partial x^2}, \qquad 0 < x < s(t) \,, \tag{4.41}$$

$$\frac{\partial \theta}{\partial t} = \frac{\partial^2 \theta}{\partial x^2}, \qquad s(t) < x \,, \tag{4.42}$$

$$[(1-c)T_I + \beta] \frac{ds}{dt} = \left. \left(k\frac{\partial T}{\partial x} - \frac{\partial \theta}{\partial x} \right) \right|_{x=s(t)} . \tag{4.43}$$

A one-phase approximation may be defined by simply neglecting the solid phase; then only the heat equation in the liquid must be solved, while the Stefan condition becomes

$$[(1-c)\, T_I(t) + \beta] \frac{ds}{dt} = - \left. \frac{\partial \theta}{\partial x} \right|_{x=s} . \tag{4.44}$$

Evans and King [46] discuss a number of papers where the Stefan problem with a variable phase change temperature is incorrectly formulated and go on to discuss in detail the approximation where (4.44) with $T_I = 0$ is used in conjunction with melting point depression (i.e. $T_I(t) < 0$). They point out that this form is popular since it arises in the case without supercooling and is accurate in the limit of large Stefan number. It may also be derived from (4.43) with the common assumption that there is no change in specific heat during the process, $c = 1$, and then choosing $k = 0$ to remove the contribution of the solid phase. Wu et al. [157] discuss similar reductions in the context of spherical nanoparticle melting. They state that when the initial temperature is different to the phase change temperature, then the one-phase limit may only be derived under the assumption $k \ll 1$.

To understand the issue, consider the total internal energy in the two-phase system. Referring to the expressions for internal energy (4.14a, 4.14b), the solid energy holds over $(0, s)$ and the liquid over (s, ∞). In the absence of curvature, the nondimensionalisation can remove the constant pressure terms, and the solid and liquid internal energies reduce to $u_s = cT$, $u_l = \theta + \beta$, where $c = c_s/c_l$. The total energy may now be written as

$$E(t) = \int_0^s cT\, dx + \int_s^\infty (\theta + \beta)\, dx\,. \tag{4.45}$$

The rate of change of energy is

$$\frac{dE}{dt} = \int_0^s c\frac{\partial T}{\partial t}\, dx + cT(s,t)\frac{ds}{dt} + \int_s^\infty \frac{\partial \theta}{\partial t}\, dx - (\theta(s,t) + \beta)\frac{ds}{dt}. \tag{4.46}$$

Replacing the time derivatives in the integrals on the right-hand side via the heat equations and integrating gives

$$\frac{dE}{dt} = k\left.\frac{\partial T}{\partial x}\right|_s - \left.\frac{\partial \theta}{\partial x}\right|_s - k\left.\frac{\partial T}{\partial x}\right|_0 + [cT(s,t) - \theta(s,t) - \beta]\frac{ds}{dt}, \tag{4.47}$$

where we have assumed $\theta_x \to 0$ as $x \to \infty$. Assuming the solid is everywhere at the interface temperature, $T(x,t) = T_I(t)$ (which varies with time), while at the interface, the liquid satisfies $\theta(s,t) = T_I(t)$ then Eq. (4.47) reduces to

$$\frac{dE}{dt} = -\left.\frac{\partial \theta}{\partial x}\right|_s - [(1-c)T_I(t) + \beta]\frac{ds}{dt}\,. \tag{4.48}$$

Energy is therefore conserved, $E_t = 0$, in this limit if the one-phase Stefan condition (4.44) is imposed. The basic one-phase Stefan condition $\beta s_t = -\,\theta_x|_{x=s}$ is also consistent with energy conservation in the limit $c \to 1$. So, it seems energy conservation should not be an issue for the one-phase reduction provided the solid is always maintained at the phase change temperature. Unfortunately, the choice $T(x,t) = T_I(t)$ is not only unrealistic but *it does not satisfy the heat equation*!

A second approach often invoked to obtain a one-phase reduction is to impose $T(x,t) = 0$, i.e. the solid temperature is now everywhere at the solidus temperature. The energy balance becomes

$$\frac{dE}{dt} = -\left.\frac{\partial \theta}{\partial x}\right|_s - [T_I(t) + \beta]\frac{ds}{dt}\,. \tag{4.49}$$

Replacing the temperature gradient via Eq. (4.44) gives

$$\frac{dE}{dt} = -cT_I s_t \, . \tag{4.50}$$

So now energy is not conserved, even in the limit $c \to 1$ where we would expect to apply the basic Stefan condition. In fact, this seems obvious on physical grounds, since there is now a jump in temperature across the interface (provided $T_I \neq 0$). On the positive side, $T(x,t) = 0$ does satisfy the heat equation.

So the naive approach of setting $T(x,t) = 0$ satisfies the solid heat equation but does not conserve energy. While setting $T(x,t) = T_I(t) \neq 0$, it is possible to conserve energy provided you don't respect the heat equation (and we should respect the heat equation, except for in Chap. 6). This problem will be resolved in the following section.

4.5.1 *Limit of Large Conductivity Ratio, k* $\gg$ 1

Heat conduction occurs on the microscopic scale due to the transfer of kinetic energy from hot, rapidly vibrating atoms or molecules to their cooler, more slowly vibrating neighbours. In solids, the close, fixed arrangement of atoms means that conduction is more efficient than in fluids, which have a larger distance between atoms. Conduction is even worse in the loose molecular arrangement of a gas. Consequently, in general, the conductivity of a solid is greater than that of its corresponding liquid phase, which is in turn greater than in the gaseous phase, for example, with ice and water $k = k_s/k_l \approx 4$, for solid and molten gold $k \approx 3$. Noting that $k > 1$, Myers and Font [106, 116] simplified the Stefan problem based on a large k reduction.

Consider a material with $k \gg 1$. The phase change is driven by some energy flux at the boundary $x = 0$, $kT_x(0,t) = -Q$, at the interface between the two phases $T(s,t) = T_I(t)$. Since cooling is not actually required to solidify a supercooled material (just a small input of energy, even a slight movement of the container, can start the process), in the literature, comparisons between one-phase and two-phase approximations have often been made by imposing zero gradient in the solid at the interface, $T_x(s,t) = 0$; see [32, 116]. Here, we examine a more general model with a nonzero heat flux at $x = 0$, which results in a nonzero gradient at the interface.

With a large k, a perturbation expansion in terms of $1/k \ll 1$ is possible. To clarify the role of k, we write the heat equation in the solid as

$$\frac{\partial T}{\partial t} = \frac{k}{c}\frac{\partial^2 T}{\partial x^2} \, , \tag{4.51}$$

that is, we have written $\alpha = k/c$ so that the role of k is apparent. Writing $T(x,t) = T_0(x,t) + T_1(x,t)/k + \cdots$, the boundary condition becomes

$$k\left.\frac{\partial T}{\partial x}\right|_{x=0} = -Q\,, \quad \Rightarrow \quad \left.\frac{\partial T_0}{\partial x}\right|_{x=0} + \frac{1}{k}\frac{\partial T_1}{\partial x} + \cdots = -\left.\frac{Q}{k}\right|_{x=0}\,, \tag{4.52}$$

and to first order in $1/k$, we obtain

$$\frac{\partial^2 T_0}{\partial x^2} = 0\,, \qquad T_0(s,t) = T_I(t)\,, \qquad \left.\frac{\partial T_0}{\partial x}\right|_{x=0} = 0 \tag{4.53}$$

$$\frac{\partial^2 T_1}{\partial x^2} = c\frac{\partial T_0}{\partial t}\,, \qquad T_1(s,t) = 0\,, \qquad \left.\frac{\partial T_1}{\partial x}\right|_{x=0} = -Q\,. \tag{4.54}$$

This leads to

$$T(x,t) = T_I(t) + \frac{1}{k}\left[c\frac{\partial T_I}{\partial t}\frac{(x^2-s^2)}{2} - Q(x-s)\right] + \mathcal{O}((1/k)^2)\,. \tag{4.55}$$

Using this to determine $T_x(s,t)$ in the two-phase Stefan condition (4.43) leads to

$$-cs\frac{\partial T_I}{\partial t} + Q + [(1-c)T_I + \beta]\,s_t = -\left.\frac{\partial \theta}{\partial x}\right|_{x=s}\,. \tag{4.56}$$

This is the appropriate leading order Stefan condition to be employed in one-phase problems with a variable interface temperature: it should be preferred over the form (4.44). It is apparent why Eq. (4.44) has been employed in the past – since the leading order temperature of (4.56) is $T_I(t)$, it is tempting, but incorrect, to assume that the leading order solid temperature contribution to the Stefan condition is zero, $T_x = 0$, and then the two-phase Stefan condition (4.43) does indeed reduce to Eq. (4.44). However, the solid temperature's contribution to the Stefan condition appears in the form kT_x, which means that the first-order term of the expansion enters at leading order, so bringing in the T_I, Q terms. The most significant difference between using (4.44) and (4.56) is the appearance of the time derivative of T_I, which can cause significant changes in the behaviour of the solution. For example, in the study of solidification with linear undercooling (or supercooling) of [116], the interface temperature varies with the velocity, $T_I = -\phi s_t$, and so the Stefan condition becomes second order in time, rather than the usual first-order equation.

To verify that the new form conserves energy, we may substitute for T via Eq. (4.55) and $\theta_x(s,t)$ via (4.56) into the energy equation (4.47) to find

$$\frac{\partial E}{\partial t} = Q\,. \tag{4.57}$$

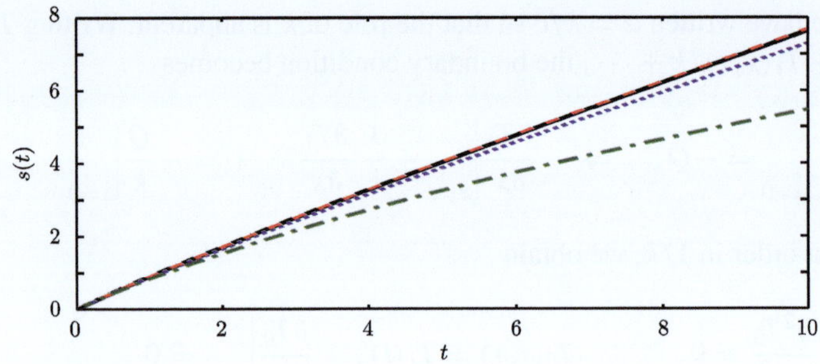

Fig. 4.2 Variation of the phase change front with time for supercooled water $k \approx 4, c \approx 0.49$ and $\beta = 0.7$. The solid black line represents the solution of the two-phase model, the dashed red line the solution of the large k model, dotted purple line the standard (non-energy conserving) one-phase reduction, and the dash-dot green line the small k model

This equation states that the rate of change of energy balances the energy input at the boundary, so this formulation conserves energy. The heat equation is also satisfied by the form of $T(x, t)$ given by Eq. (4.55).

We note that in [46], a small k approximation is derived:

$$(T_I + \beta)\frac{\partial s}{\partial t} = -\left.\frac{\partial \theta}{\partial x}\right|_{x=s} . \tag{4.58}$$

This may be retrieved from (4.43) by neglecting the solid components, i.e. by setting $c = k = 0$.

In Fig. 4.2, we compare the evolution of the phase change front for the current large k model (4.56) (dashed line), small k model (4.58) (dash-dot), standard one-phase model (4.44) (dotted), and the two-phase model (4.43) (solid line). Parameter values appropriate to water were used $k \approx 4, c \approx 0.49$ and $\beta = 0.7$. This β value corresponds to an undercooling of 115 K below the bulk phase change temperature of 273.15 K – in practice, it is hard to pass below 48 K undercooling, which is the homogeneous nucleation temperature (the temperature at which nucleation can occur in the bulk without a nucleation site). These curves were computed numerically using the boundary immobilisation method and Keller box finite difference technique; see [92, 116]. They have also been verified using the TIM, with almost identical results (feel free to try it at home); see [52]. As can be seen, the large k result is almost identical to the two-phase solution. The standard reduction, which does not conserve energy, is surprisingly accurate, while the small k solution, which does conserve energy, does not capture the behaviour well.

To demonstrate the fact that the supercooled model can produce solutions for both $\beta < 1$ and $\beta > 1$ in Fig. 4.3, we show results for the case where $\beta = 1.3$ (a supercooling of 62 K). The results are similar to the previous case, with the large k model proving highly accurate and the reduction that loses energy providing reasonable agreement. The worst is the energy-conserving small k approximation.

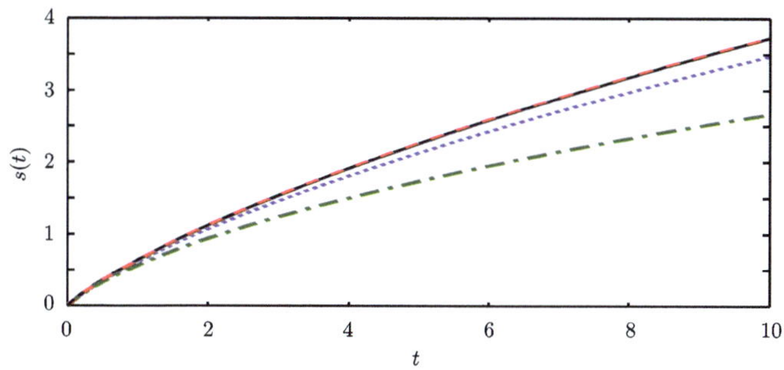

Fig. 4.3 Variation of the phase change front with time for supercooled water $k \approx 4$, $c \approx 0.49$ and $\beta = 1.3$. The solid black line represents the solution of the two-phase model, the red dashed line the solution of the large k model, purple dotted line the standard (non-energy conserving) one-phase reduction, and the green dash-dot line the small k model

If we continue to increase β, all curves tend towards the two-phase solution and, for sufficiently large β, even the widely used non-energy-conserving form will give accurate solutions.

Results are also presented in [116] for the solidification of salol again for $\beta = 0.7, 1.3$ (but now corresponding to supercooling of 80, 42 K, respectively, below the phase change temperature of 316 K). Salol was chosen since it is a material with a very low value of $k \approx 1.4$. Despite this low value, suggesting the large k reduction will be inaccurate, the results take almost exactly the same form as those for water with the large k result being, by far, the best approximation to the two-phase result.

In Font et al. [53], the large k reduction for cylindrical and spherically symmetric systems is developed. Results are compared for the melting of gold nanospheres and nanowires. Again, the large k reduction is highly accurate.

4.6 Conclusions

The governing equations describing the phase change of a material with distinct thermophysical properties in each phase were derived from conservation laws. This is a natural approach for a two-phase mixture, with a finite region over which properties change. As the width of this region approaches zero, the sharp-interface limit, used in many Stefan problems, is retrieved. The one drawback to this approach is that being the limit of a mixture model, there is no mechanism to account for surface tension. This could be dealt with formally by introducing an appropriate Korteweg stress tensor, but here we have added a source term at the interface. From the conservation laws, the standard heat equations are obtained. Applying the Rankine-Hugoniot condition to the conservation laws leads to relations between the material and interface velocities and the pressure and energy jump. The latter expression provides the Stefan condition.

An important point to note is that the obtained Stefan condition differs to those generally found in the literature. Certain differences relate to the focus of previous authors; however, a particular error concerns the kinetic energy contribution that has propagated in the literature. The Stefan condition derived here clarifies the various energetic contributions and is written in such a way that it may be readily incorporated into non-Fourier-based studies as well.

A further issue in the literature has been the use of one-phase models with a variable phase change temperature. Naively dropping the effect of one of the phases results in a system that does not conserve energy. A simple perturbation analysis, in the limit of large conductivity ratio $k_s \gg k_l$, showed that the problem is easily rectified through the addition of a term to the Stefan condition that depends on the interface temperature. With this term, the one-phase model provides results that are almost identical to the two-phase model and conserves energy.

To illustrate the use of the governing equations, the system was adapted to the standard problems of solidification of a semi-infinite fluid and melting of a sphere (to be honest, these examples were chosen for later use). In subsequent chapters, we will exploit these reductions as well as other cases when dealing with the melting of nanoparticles and wires, crystal growth, and the solidification of a supercooled liquid.

Chapter 4 Exercises

1. Assuming incompressible materials, a stationary solid and zero surface tension apply the Rankine-Hugoniot condition to the equation for conservation of energy (4.1c) to determine a relation between the fluid velocity and pressure.

   If $\rho_s > \rho_l$, as is the case for most materials, what does the sign of the pressure indicate?
2. Equation (4.20) defines the generalised Stefan condition including interfacial temperature variation and kinetic energy:

$$\rho_s \left[L_m + (c_l - c_s)(T_I - T_m) + \frac{1}{2}\left(\left(\frac{\rho_s}{\rho_l}\right)^2 - 1\right)\left(\frac{\partial s}{\partial t}\right)^2 \right] \frac{\partial s}{\partial t}$$

$$= k_s \nabla T \cdot \hat{\mathbf{n}} - k_l \nabla \theta \cdot \hat{\mathbf{n}} \,,$$

   Assume $s = 2\lambda\sqrt{t}$ provides an approximate solution. For a supercooling of 10 K and a water-ice system, determine the time when kinetic energy is comparable to the sensible heat and latent heat. Appropriate parameter values are provided in Table 1.1, and from Sect. 2.1.1, we have $\lambda = 0.143$.

   Given the sizes you find, does it make sense to ever include these terms?
3. Following Sect. 4.3.2, determine the liquid velocity for the symmetric melting of a cylindrical material and hence the position of the outer boundary $R_b(t)$ in terms of $R(t)$.

4. In [49], the melting of a nanowire due to a constant temperature thermal bath is studied. The following system is defined:

$$\frac{1}{\beta}\frac{\partial T}{\partial t} = \frac{1}{r}\frac{\partial}{\partial r}\left(r\frac{\partial T}{\partial r}\right), \qquad 0 < r < R(t) ,$$

$$\frac{1}{\beta}\frac{\partial \theta}{\partial t} = \frac{1}{r}\frac{\partial}{\partial r}\left(r\frac{\partial \theta}{\partial r}\right), \qquad R(t) < r < 1 ,$$

$$\left[1 - \frac{\Lambda}{2R\beta}\right]\frac{dR}{dt} = \left(k\frac{\partial T}{\partial r} - \frac{\partial \theta}{\partial r}\right)\bigg|_{r=R(t)} ,$$

subject to

$$\theta(1,t) = 1 , \quad \frac{\partial T}{\partial r}\bigg|_{r=0} = 0 , \quad \theta(R,t) = T(R,t) = -\frac{\Gamma}{2R}$$

where Λ, Γ are constants representing the effect of surface tension. This is based on the assumption of perfect thermal contact at the outer boundary; no density change between phases and the melt temperature is proportional to curvature and therefore decreases with the radius (this is a form of Gibbs-Thomson relation, discussed in the following chapter):

(a) By taking the leading order terms of a perturbation expansion in $1/\beta \ll 1$, determine the temperature profile in both solid and liquid regions.
(b) From the temperature expressions, derive an equation for the rate of change of the radius, R_t. Note, there are two periods when $R_t \to \infty$; when does this happen? Can you provide a physical explanation?
(c) When the Gibbs-Thomson effect is negligible, $\Gamma = 0$ (also Λ, but this only affects the first-order solution), the system reduces to a standard form of Stefan problem. In this case, determine the solution for $t(R)$. One of the most important and simple checks on a solution is to plot it. Does the solution make physical sense? Does it satisfy the boundary/initial conditions? Plot your solution (for $R \in (0, 1)$)!

5. In [49], a second case is considered, that of solidification of a cylindrical liquid core. The model is virtually identical to the one specified above, except due to the change in the direction of the normal, the melt temperature is $\Gamma/(2R)$ and hence

$$\left[1 + \frac{\Lambda}{2R\beta}\right]\frac{dR}{dt} = \left(k\frac{\partial T}{\partial r} - \frac{\partial \theta}{\partial r}\right)\bigg|_{r=R(t)} .$$

The boundary conditions become

$$T(1,t) = 1 , \quad \frac{\partial \theta}{\partial r}\bigg|_{r=0} = 0 , \quad \theta(R,t) = T(R,t) = \frac{\Gamma}{2R}$$

(a) Without resorting to any mathematics, will solidification be faster or slower than the previous melting problem?
(b) Now solve the leading order problem to determine R_t and hence provide a mathematical answer to the question of solidification versus melting rate.

6. Verify energy conservation, defined by Eq. (4.57) through substituting for the temperature $T(x,t)$ of (4.55) and using the Stefan condition (4.56) to determine θ_x.
7. Consider the system defined by Eqs. (4.41–4.43) and naively drop the solid contribution such that

$$\frac{\partial \theta}{\partial t} = \frac{\partial^2 \theta}{\partial x^2}\,, \qquad [T_I + \beta]\frac{ds}{dt} = -\frac{\partial \theta}{\partial x}\bigg|_{x=s(t)}\,.$$

These hold over the domain $x > s(t)$, where $s(0) = 0$ and are subject to $\theta \to -1$ and $x \to \infty$, $\theta(s,t) = T_I$, $\theta(x,0) = -1$. This may be considered to represent a supercooled liquid that freezes due to some initial perturbation at the origin:

(a) If $T_I(t) = 0$, solve the system using the similarity variable $\eta = x/(2\sqrt{t})$. Define the transcendental equation for λ, where $s = 2\lambda\sqrt{t}$. What happens as $\beta \to 1$ and what is the physical significance?
(b) Calculate the rate of change of energy. Does this system conserve energy?
(c) Now consider the supercooled system, such that $T_I(t) = -s_t$. In this case, we cannot use the similarity solution, but even without this, it is possible to derive the expression for rate of change of energy. Does this system conserve energy?

8. Consider a melting sphere with outer radius $R_b(t)$ and interface $R(t)$. The total energy in the system is denoted by

$$\frac{E}{4\pi} = \int_0^R \rho_s u_s r^2\,dr + \int_R^{R_b} \rho_l\left(u_l + \frac{v_l^2}{2}\right) r^2\,dr + R^2\sigma_{sl} + R_b^2\sigma_{la}\,.$$

This simply states that the energy incorporates internal energy of the solid and liquid as well as surface energy at the two interfaces:

(a) Take the derivative with respect to time to demonstrate that

$$\begin{aligned}\frac{1}{4\pi}\frac{dE}{dt} &= R^2\left[q_l - q_s - \left\{\rho_s\left(u_l - u_s + \frac{v_l^2}{2}\right) - \sigma_{sl}\kappa\right.\right.\\ &\left.\left.\quad - \left(1 - \frac{\rho_s}{\rho_l}\right)p_l\right\}\frac{dR}{dt}\right]_{r=R}\\ &\quad - R_b^2\left[q_l + (p_l - \sigma_{la}\kappa)\frac{dR_b}{dt}\right]_{r=R_b}\,.\end{aligned}$$

(b) Reduce the above equation by invoking equation (4.13) to demonstrate that energy change is driven solely by the heat flow and pressure work at the outer boundary.

(c) The sphere must do work against the atmospheric pressure – the associated rate of energy change on the atmosphere side is

$$\frac{1}{4\pi}\frac{dE_a}{dt} = -R_b^2 p_a \; v_l|_{r=R_b}$$

Thermal energy is transferred across the boundary proportional to the temperature difference:

$$\frac{1}{4\pi}\frac{dE_n}{dt} = R_b^2 \mathscr{H}_{la}(T_a - \theta(R_b, t)) .$$

The jump in energy across the boundary is then given by the energy balance:

$$E - E_a = E_n .$$

From this, derive the appropriate form of boundary condition at $r = R_b$ (note, you will have to use the Young-Laplace equation $p_l - p_a = \sigma_{la}\kappa$ and recall $v_l(R_b, t) = \partial R_b/\partial t$)

Chapter 5
Phase Change with a Variable Interface Temperature

The vast majority of investigations into Stefan problems, both theoretical and experimental, have dealt with a constant phase change temperature. However, there are many physically important applications where this temperature changes from its standard value. Superheating is the phenomenon where a liquid is heated above its boiling point, without boiling. Supercooling (or undercooling) is when the liquid is cooled below the freezing temperature, without freezing. For both boiling and freezing, this may be achieved through the removal of nucleation sites, for example, by using purified liquids and very smooth containers. However, these states are not stable; they are termed metastable [42], and any disturbance is likely to result in a rapid, almost violent, phase change.

A supercooled liquid can solidify much more rapidly than a non-supercooled one, and when such rapid solidification occurs, the liquid may not have time to form its usual crystalline structure. Materials made from supercooled melts can therefore have markedly different properties to the standard form of the material. A material formed from a supercooled liquid, usually called a glassy or amorphous solid, can present greater corrosion resistance, toughness, strength, hardness, and elasticity than common materials: amorphous metal alloys can be twice as strong and three times more elastic than steel. Such materials are currently used in medicine, defence and aerospace equipment, electronics, 3D printing, and sports; see [41, 51].

A common strategy to reduce the freezing temperature is to increase the ambient pressure. If the pressure is subsequently released, almost instantaneous freezing occurs. This is exploited in the food industry and permits the freezing of certain products that normally spoil when frozen more slowly. It is also used in cryopreservation. Pressure may vary due to surface tension effects at a curved interface. Hence, freezing fronts with high curvature tend to exhibit a variable phase change temperature.

In this chapter, we investigate Stefan problems where the phase change temperature differs from the standard value. Two situations are considered. The first concerns the solidification of a supercooled or undercooled liquid; we will use both

T. G. Myers, *Theoretical and Practical Stefan Problems*, Interdisciplinary Applied Mathematics 63, https://doi.org/10.1007/978-3-032-04826-4_5

terms in the following (but note that while it is common to use the term supercooled liquid, when speaking of the temperature below the standard phase change value $T_m - T_I$, this is more commonly termed the level of undercooling). The second case concerns the phase change of nanoparticles and nanowires. These involve high curvatures and the resultant stress acts to alter the phase change temperature. A third problem will also be discussed concerning the growth of a nanocrystal in a monomer solution. This is a mass diffusion rather than heat diffusion process; however, it is mathematically analogous. It is treated in this section since the "solubility" of the crystal, its ability to take on or lose new material, is proportional to its size (the solubility equates to the phase change temperature), and so the model corresponds to a Stefan problem with variable phase change temperature.

The chapter closely follows work described in Myers and Fanelli [105] and Fanelli et al. [47] on nanocrystal growth; Font and Myers [52] on solidification of supercooled melts; Florio and Myers [49] on nanowire melting; and Myers, Font, Mitchell, and Ribera [51, 53, 102, 106, 132] on nanoparticle phase change. Other notable theoretical studies in the field of nanoparticle melting include the work of McCue et al. [9, 10, 82, 157, 158].

5.1 Models for the Interface Temperature

Before leaping straight into a new set of Stefan problems, we will take a brief aside to look at how and why the phase change temperature may vary.

Ashby and Jones [7] use a statistical mechanics argument to demonstrate that in the presence of supercooling, the solidification rate takes the form

$$\frac{ds}{dt} = \omega \exp(-q/(k_B T_I)) \frac{(T_m - T_I)}{T_m}, \tag{5.1}$$

where q is the activation energy and $k_B \approx 1.381 \times 10^{-23}$ J/K is the Boltzmann constant. The coefficient $\omega = d\Delta H/(6h)$(m/s) where d(m) is the molecular diameter and $h \approx 6.626 \times 10^{-34}$(J s) is Planck's constant, $\Delta H = L_m M/N_A$(J) with M(kg/mol) the molecular mass and $N_A \approx 6.022 \times 10^{23}\text{mol}^{-1}$ is Avogadro's number. Implicit in the derivation is that the molecule is approximately spherical and so equally likely to move in any of the six principal directions (hence the factor one-sixth in ω) and a molecular "diameter" is well defined. Equation (5.1) results in a solidification rate that first increases with the level of supercooling $T_m - T_I$ but subsequently decreases to zero.

The argument put forward by Davis [40] to explain this nonlinear behaviour concerns a liquid at an interface with temperature T_I. If the interface temperature is the same as the standard phase change temperature or the *heterogeneous nucleation temperature* (HNT), $T_I = T_m$, then without any external input, the interface is in equilibrium; without molecules attaching and detaching from the interface at the same rate, it therefore remains stationary. As T_I reduces below T_m, the liquid molecules attach more rapidly to the interface than they detach, and the velocity s_t increases with an increase in supercooling $T_m - T_I$. However, the motion of the liquid molecules towards the interface depends on their temperature. At a certain level of supercooling, the liquid molecules have little energy, which then affects their mobility and hence their ability to move to the solid interface: the molecules simply cannot attach fast enough to the surface. At this stage, the freezing rate decreases.

A linearised version of (5.1) is often dealt with in the literature, since this lends itself to analysis. It is based on the assumption that $T_m - T_I$ is small. In this case, an explicit expression for the interface temperature may be written in terms of the velocity:

$$T_I = T_m - \phi s_t \,, \tag{5.2}$$

where $\phi = T_m \exp(q/(k_B T_m)/\omega$ [40]. Setting $\phi = 0$ in this equation brings us back to the standard condition that the interface temperature takes the constant value T_m. With $\phi > 0$, Eq. (5.2) defines a *linear kinetic undercooling* condition.

In Font et al. [52], two examples are studied: firstly, the solidification of supercooled copper which is relevant to practical applications but has a very fast solidification rate (and hence it is very difficult to take measurements), and secondly, salol, which has a very slow solidification rate and so lends itself to laboratory experiments. For copper, the molecular diameter d is well defined (since it is the diameter of the atom), while salol is a molecule ($C_{13}H_{10}O_3$) with no single diameter and presumably with preferential directions for movement. In this case, it is more reliable to choose the coefficient ω by matching to a known experimental solidification rate. In [51], the maximum solidification rate for copper is given as $s_t = 2.9$ m/s at $T_I = 1108$ K, where $T_m = 1360$ K. Rearranging equation (5.1) and using this maximum value of s_t and $q = 6.71 \times 10^{-20}$ J, we find

$$\omega = \exp(q/(k_B T_I)) \frac{T_m}{(T_m - T_I)} \frac{ds}{dt} \approx 1.257 \times 10^3 \text{ m/s} \,. \tag{5.3}$$

This is very close to the value obtained from the definition $\omega = d\Delta H/(6h) = 1.276 \times 10^3$ m/s (using the values $d = 0.23$ nm, $\Delta H = 2.2 \times 10^{-20}$ J). For salol [51] quotes a maximum rate $s_t = 4.4 \times 10^{-5}$m/s when $T_I = 298$ K, $T_m = 316$ K and $q = 6.61 \times 10^{-20}$ J, resulting in $\omega = 7.39 \times 10^3$ m/s. Other thermophysical parameters for copper and salol are provided in Table 5.1.

Table 5.1 Approximate thermodynamical parameter values for Cu and salol, taken from [7, 22, 44, 83]

Substance	T_m (K) ($\times 10^3$)	L_m (J/kg) ($\times 10^5$)	c_l (J/kg·K) ($\times 10^3$)	ρ_l (kg/m^3) ($\times 10^3$)	k_l (W/m·K)	q (J) ($\times 10^{-20}$)	d (m) ($\times 10^{-9}$)
Copper	1.36	2.09	0.53	8.02	157.18	6.71	0.23
Salol	0.316	0.90	1.58	1.18	0.18	6.61	1.00

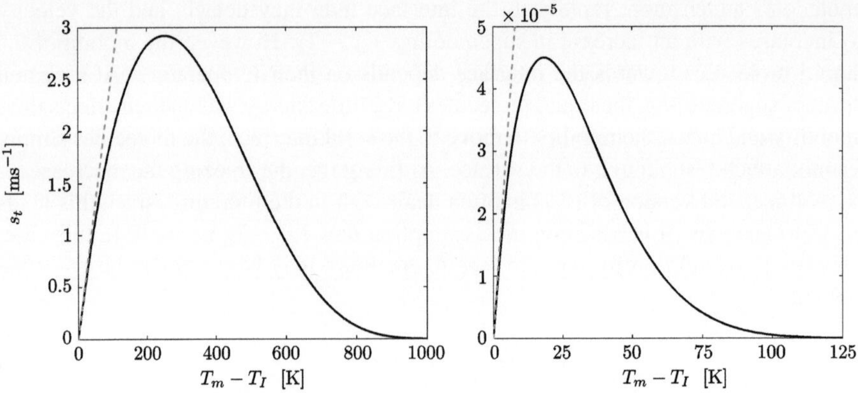

Fig. 5.1 Representation of the solidification speed of copper (left) and salol (right) as a function of the supercooling, $T_m - T_I$. The solid line represents the full expression for s_t, Eq. (5.1), and the dashed line the linear approximation, Eq. (5.2). Appropriate parameter values are provided in Table 5.1

The full nonlinear relation for the front velocity against the degree of supercooling, Eq. (5.1), is shown in Fig. 5.1 as a solid line. For copper, Fig. 5.1 (left), the form makes it clear that for a low degree of supercooling, $T_m - T_I < 250$ K, the rate of solidification increases. Beyond this, as the molecules become sluggish, the rate decreases such that at around 1000 K of supercooling, solidification is negligible. The dashed line represents the linear approximation, Eq. (5.2), which is really only a good approximation for $T_m - T_I$ <30 K. In Fig. 5.1 (right), we show the equivalent result for salol, which peaks at $s_t = 4.4 \times 10^{-5}$ m/s at around 25 K supercooling.

As mentioned earlier, another mechanism to produce a variation in the phase change temperature is through changing the ambient pressure. This could be through a mechanical process or, in the case of liquids with high curvature, via surface tension. At a highly curved surface, the surface tension-induced stress may be sufficient to significantly affect the pressure and hence phase change temperature. This has been considered in detail in the theoretical studies of nanoparticle melting [51, 82, 132, 157] and nanowire melting [49]. Experimentally, Buffat and Borel [19] reported a decrease of approximately 500 K below the bulk melt temperature (approximately 60%) for gold nanoparticles with radii a little above 1 nm. Decreases of 70 and 200 K have been reported for tin and lead nanoparticles [39]. The most commonly used form to describe the decrease of phase change temperature with curvature is the Gibbs-Thomson relation:

$$T_I = T_m \left(1 - \frac{\sigma_{sl} \kappa}{\rho_s L_m}\right) , \tag{5.4}$$

where κ is the sum of principal curvatures (for a sphere of radius R, $\kappa = 2/R$; for a cylinder, $\kappa = 1/R$). This is derived via variational calculus in [40]. Font and Myers

[51] state that it is a particular limit of a more generalised Gibbs-Thomson relation:

$$\begin{aligned}\left(\frac{1}{\rho_l}-\frac{1}{\rho_s}\right)(p_l-p_s)=&L_m\left(\frac{T_I}{T_m}-1\right)\\&+(c_l-c_s)\left[T_I\ln\left(\frac{T_I}{T_m}\right)+T_m-T_I\right]+\frac{\sigma_{sl}\kappa}{\rho_s},\end{aligned} \tag{5.5}$$

which is derived from thermodynamical principles in [4]. They go on to show that the pressure term is typically two orders of magnitude smaller than the next smallest term and so may be neglected. Excellent agreement was then demonstrated between the prediction of the reduced (5.5) (neglecting pressure) and experimental data for the observed melt temperature of gold nanoparticles down to a radius of 1 nm [51, Fig. 1]. They go onto show that for radii below approximately 1 nm, where $T_m \approx 329$ K, Eq. (5.5), becomes multivalued, suggesting that it should not be applied below this value.

5.2 The Solidification of a Supercooled Liquid

Here we follow the work of Font et al. [52], which employs a powerful combination of the analytical techniques described in Chap. 2 to shed light on this complex problem. The methods employed include similarity reductions, travelling waves, boundary fixing, perturbations, and small and large time solutions. In the second half of their paper, these techniques are used again in combination with the opTimal Integral Method (TIM); see Sect. 2.2.4. Below we will use the first part of the paper, without the TIM analysis, since this provides all the important results. However, we will show results from the TIM analysis to demonstrate its accuracy and also because in the final example, with nonlinear supercooling (undercooling), there is no analytical expression.

In Font et al. [52], the one-dimensional solidification of a supercooled liquid initially occupying $x \geq 0$ was considered. The solidification of a supercooled liquid does not need a cold boundary: a nucleation site or even just some input of energy (such as a kick to the container) can initiate the process. So for this study, we assume the solid phase is always close to the homogeneous nucleation temperature and study the following one-phase Stefan problem:

$$\frac{\partial\theta}{\partial t}=\alpha_l\frac{\partial^2\theta}{\partial x^2}, \qquad s<x<\infty \tag{5.6a}$$

$$\theta=T_I(t), \qquad \rho_l L_m s_t=-k_l\frac{\partial\theta}{\partial x}, \qquad \text{at } x=s \tag{5.6b}$$

$$\theta\to T_\infty, \qquad \text{as } x\to\infty, \tag{5.6c}$$

$$\theta(x,0)=T_\infty, \qquad s(0)=0, \tag{5.6d}$$

where $T_\infty < T_m$. The interface temperature is related to the interface velocity by Eq. (5.1); the linearised version is given by (5.2).

The system (5.6a)–(5.6d) may be nondimensionalised in the standard fashion:

$$\hat{\theta} = \frac{\theta - T_m}{\Delta T}, \qquad \hat{t} = \frac{t}{\tau}, \qquad \hat{x} = \frac{x}{L}, \qquad \hat{s} = \frac{s}{L}, \qquad \hat{T}_I = \frac{T_I - T_m}{\Delta T} < 0, \tag{5.7}$$

where $\tau = L^2/\alpha_l$, $\Delta T = T_m - T_\infty$, and $L = \phi\alpha_l/\Delta T$ (chosen to balance terms in Eq. (5.2)). Note, in the case where $T_I(t) = T_m$, there is no natural length-scale and it must be left arbitrary. Dropping the hat notation, the governing equations are now

$$\frac{\partial \theta}{\partial t} = \frac{\partial^2 \theta}{\partial x^2}, \qquad s < x < \infty \tag{5.8a}$$

$$\theta = T_I(t), \qquad \beta s_t = -\frac{\partial \theta}{\partial x}, \qquad \text{at } x = s \tag{5.8b}$$

$$\theta \to -1, \qquad \text{as } x \to \infty \tag{5.8c}$$

$$\theta(x, 0) = -1, \qquad s(0) = 0, \tag{5.8d}$$

where $\beta = L_m/(c_l \Delta T)$ and T_I now refers to the nondimensional interface temperature.

An interesting feature of this system is that in the case where the interface temperature is fixed $T_I(t) = 0$ (i.e. the heterogeneous nucleation temperature) and the liquid is initially everywhere below this $\theta(x, 0) < 0$, then solutions only exist for $\beta > 1$. As $\beta \to 1^+$, the interface speed tends to infinity. This demonstrates the necessity of supercooling relations such as (5.1) or (5.2). Applying these relations, once the interface temperature is allowed to decrease below the HNT, solutions may be obtained for any $\beta > 0$. For the case of linear kinetic undercooling, where the interface temperature follows (5.2), Davis [40] shows that for $\beta > 1$, $s \propto t^{1/2}$ (as we have seen from the Neumann solution of Chap. 2), while for $\beta < 1$, $s \propto t$ (satisfying a travelling wave). For $\beta = 1$, Davis refers the reader to a Russian paper that proves $s \propto t^{2/3}$. We will see later why he wisely omits this final proof.

As discussed in Sect. 4.5, the system (5.6) (or (5.8)) in fact loses energy. However, as we have demonstrated, this form gives a reasonable approximation and is accurate for sufficiently large β. Since (5.8a)–(5.8d) is the form studied in [40] (and references therein) and [52], we retain it for the following analysis. This allows us to frame the results in the context of previous work but also provides an opportunity for future researchers to extend the analysis using the more complete Stefan condition (4.56).

The interface temperature T_I is now defined through the nondimensional version of Eq. (5.1):

$$s_t = -T_I \exp\left(\frac{QT_I}{P + T_I}\right), \tag{5.9}$$

where $Q = q/(kT_m)$ and $P = T_m/\Delta T$. The case of no supercooling is retrieved by setting $T_I(t) = 0$ in (5.8b) and then (5.9) is not used. Small supercooling requires ΔT to be small such that $P \to \infty$, and the exponential term of (5.9) may be set to unity, and then $s_t = -T_I(t)$. Otherwise, the full expression in (5.9) is used to determine T_I.

When $T_I(t)$ is not identically zero, then we require an initial value (and also a value for $s_t(0)$). Noting that initially $\theta(x, 0) = -1$ while at the interface $T_I(t) = \theta(s, t)$, then $s(0) = 0$ indicates that $T_I(0) = \theta(0, 0) = -1$ matches the initial temperature. An initial (finite!) value for s_t may then be found by substituting $T_I(0) = -1$ into Eq. (5.9) or, with the linearised version $s_t(0) = 1$. In the case $T_I \equiv 0$, then as $t \to 0$, the discontinuity in temperature due to the conflicting conditions $\theta(0, 0) = 0, \theta(x, 0) = -1$ cannot be avoided. This indicates that the front velocity $s_t = -\theta_x(s, t)$ is infinite at $t = 0$, as was noted in previous chapters.

We begin by fixing the boundary using one of the transformations introduced in Sect. 2.2.1:

$$y = x - s\,, \qquad \theta(x, t) = F(y, t)\,. \tag{5.10}$$

Then the system (5.8) becomes

$$\frac{\partial^2 F}{\partial y^2} = \frac{\partial F}{\partial t} - s_t \frac{\partial F}{\partial y}\,, \qquad 0 < y < \infty\,, \tag{5.11a}$$

$$F = T_I(t)\,, \quad \beta s_t = -\frac{\partial F}{\partial y}\,, \qquad \text{at } y = 0 \tag{5.11b}$$

$$F \to -1\,, \qquad \text{as } y \to \infty \tag{5.11c}$$

$$F(y, 0) = -1\,, \qquad s(0) = 0\,. \tag{5.11d}$$

5.2.1 Asymptotic Analysis

Most exact solutions to Stefan problems can be obtained by similarity methods. Hence, we begin by seeking similarity solutions to the system (5.11) and so define

$$\eta = \frac{y}{t^\alpha}\,, \qquad F(y, t) = G(\eta)\,. \tag{5.12}$$

The new co-ordinate may be viewed as a form of Landau transformation, Eq. (2.49), if $s \propto t^{\alpha}$. Under this transformation, the governing equations (5.11a)–(5.11c) become

$$G'' = -\left(\alpha\eta t^{2\alpha-1} + t^{\alpha}s_t\right)G', \qquad 0 < \eta < \infty \tag{5.13}$$

$$G(0) = T_I(t), \qquad G|_{\eta\to\infty} \to -1, \qquad \beta t^{\alpha}s_t = -G'(0), \tag{5.14}$$

with $s(0) = 0$, and primes denote differentiation with respect to η. Defining the constant $\nu = -G'(0)/\beta \geq 0$, we use the Stefan condition to eliminate $t^{\alpha}s_t$ from (5.13) and so the model becomes

$$G'' = -\left(\alpha\eta t^{2\alpha-1} + \nu\right)G', \qquad 0 < \eta < \infty \tag{5.15}$$

$$G(0) = T_I(t), \qquad G|_{\eta\to\infty} \to -1, \qquad t^{\alpha}s_t = \nu, \qquad s(0) = 0. \tag{5.16}$$

Since $G = G(\eta)$, this system is strictly only valid when the t dependence is removed from the equations involving G. In (5.15), this may be achieved by setting $\alpha = 0$ or $1/2$, but we must also take into account the time-dependent boundary condition $G(0) = T_I(t)$.

5.2.2 No Kinetic Undercooling

The classical formulation for the one-phase supercooled Stefan problem with no kinetic undercooling is obtained by setting $T_I(t) = 0$. Following standard analysis, we choose $\alpha = 1/2$, and so Eqs. (5.15, 5.16) become

$$G'' = -\left(\frac{\eta}{2} + \nu\right)G', \qquad 0 < \eta < \infty \tag{5.17}$$

$$G(0) = 0, \quad t^{1/2}s_t = \nu, \qquad G|_{\eta\to\infty} \to -1. \tag{5.18}$$

We encountered equations of the form (5.17) in Chap. 2 when dealing with Neumann's solution and so are familiar with their integration, the result being

$$G(\eta) = -1 + \frac{\operatorname{erfc}(\eta/2 + \nu))}{\operatorname{erfc}(\nu)}, \qquad s = 2\nu\sqrt{t}. \tag{5.19}$$

Taking the derivative of $G(\eta)$ and using the definition of $\nu = -G'(0)/\beta$, we obtain the transcendental equation:

$$\beta\sqrt{\pi}\nu\operatorname{erfc}(\nu)e^{\nu^2} = 1. \tag{5.20}$$

In the original variables, the solution is

$$\theta = -1 + \frac{\operatorname{erfc}(x/(2\sqrt{t}))}{\operatorname{erfc}(\nu)} , \qquad s = 2\nu\sqrt{t} , \tag{5.21}$$

where $\nu(\beta)$ satisfies (5.20). This is a slight modification of the one-phase Neumann solution (2.30), where instead of the solid temperature for $x \leq s$, we now have the liquid temperature for $x \geq s$. The expression for $s(t)$ demonstrates that this solution has infinite velocity at $t = 0$ (since $s_t \propto 1/\sqrt{t}$). Equation (5.20) is dependent on $\beta = L_m/(c_l \Delta T)$ and determines $\nu(\beta)$; it has a unique solution for $\beta > 1$, but when $\beta \leq 1$, no solution exists; as $\beta \to 1^+$, $\nu \to \infty$ and the front velocity is ∞ for all time. This is very different to the examples studied in earlier sections, where solutions could be obtained for any value of β.

5.2.3 *Linear Kinetic Undercooling*

5.2.3.1 Small Time Solutions

As mentioned repeatedly, in the absence of kinetic undercooling, the fixed boundary temperature condition leads to an initial infinite temperature gradient. Including kinetic undercooling, $\theta(0, 0) = T_I(0) \approx -1$, permits a finite temperature gradient. This indicates that at small times, solutions with and without kinetic undercooling have very different early time behaviour and so require different scalings at small times.

Let us now consider the governing equations (5.11). Noting that $F(y, 0) = -1$, we look for a small time similarity solution close to $F = -1$ and so re-define the function $G(\eta)$ via

$$\eta = \frac{y}{t^{1/2}} , \qquad F(y, t) = -1 + t^{1/2} G(\eta) . \tag{5.22}$$

This transforms equations (5.11) into

$$G'' = \frac{1}{2} G - \frac{\eta}{2} G' - t^{1/2} s_t G' , \qquad 0 < \eta < \infty \tag{5.23}$$

$$-1 + t^{1/2} G(0) = -s_t , \quad \beta s_t = -G'(0) , \qquad G|_{\eta \to \infty} \to 0 . \tag{5.24}$$

In all the cases of this section, we aim to balance the two terms in the Stefan condition, to ensure motion at leading order. Here the factor $t^{1/2}$ in front of $G(\eta)$ in (5.22) was chosen to provide a balance in the Stefan condition of (5.24), $\beta s_t = -G'(0)$. From this equation, since $G'(0)$, β are constant, it follows that $s \propto t$, but this choice has left time in the governing equation (5.23) and the boundary condition (5.24a): for this reason, we make the small time substitution $t = \varepsilon\tau$, $s = \varepsilon r$, where

ε is an artificial small parameter [31, 72], to find

$$G'' = \frac{1}{2}G - \frac{\eta}{2}G' - \varepsilon^{1/2}\tau^{1/2}r_\tau G' , \qquad 0 < \eta < \infty \tag{5.25}$$

$$-1 + \varepsilon^{1/2}\tau^{1/2}G(0) = -r_\tau , \quad r_\tau = \nu , \qquad G|_{\eta\to\infty} \to 0 , \tag{5.26}$$

where again $\nu = -G'(0)/\beta$. This form motivates a perturbation series with $G(\eta) = G_0(\eta)+\varepsilon^{1/2}G_1(\eta)+\mathcal{O}(\varepsilon)$ and $r(\tau) = r_0(\tau)+\varepsilon^{1/2}r_1(\tau)+\mathcal{O}(\varepsilon)$, so that the leading order form for G is independent of τ. The leading order problem is then

$$G_0'' = \frac{1}{2}G_0 - \frac{\eta}{2}G_0' , \qquad 0 < \eta < \infty \tag{5.27}$$

$$1 = r_{0\tau} , \quad r_{0\tau} = \nu , \qquad G_0|_{\eta\to\infty} = 0 . \tag{5.28}$$

Combining the first two boundary conditions in (5.28) shows that $\nu = 1$, and this provides the second boundary condition for G_0, namely, $G_0'(0) = -\beta$ (since by definition $\nu = -G'(0)/\beta$). The solution is then

$$G_0(\eta) = \beta\left[\frac{2}{\sqrt{\pi}}e^{-\eta^2/4} - \eta\,\mathrm{erfc}\left(\frac{\eta}{2}\right)\right] , \qquad r_0(\tau) = \tau . \tag{5.29}$$

The self-similarity assumption in (5.22) does not permit us to solve for the temperature at first order (due to the presence of τ in Eq. (5.25)); however, we may look for the first-order solution of $r(\tau)$ through Eq. (5.26a):

$$r_{1\tau} = -\tau^{1/2}G_0(0) \qquad \Rightarrow \qquad r_1 = -\frac{4\beta\tau^{3/2}}{3\sqrt{\pi}} ,$$

where $G_0(0) = 2\beta/\sqrt{\pi}$. Writing this solution back in the original variables gives

$$\theta(x,t) \approx -1 + \beta\left[2\sqrt{\frac{t}{\pi}}\exp\left(-\frac{(x-s)^2}{4t}\right) - (x-s)\mathrm{erfc}\left(\frac{x-s}{2\sqrt{t}}\right)\right] , \tag{5.30}$$

$$s \approx t\left[1 - \frac{4\beta}{3\sqrt{\pi}}\sqrt{t}\right] , \qquad s_t \approx 1 - \frac{2\beta}{\sqrt{\pi}}\sqrt{t} . \tag{5.31}$$

In contrast to the previous front velocity, here the velocity is constant as $t \to 0$. Recall that the temperature is accurate only to leading order, while the position of the front is accurate to first order. This small time solution is valid for any $\beta > 0$.

5.2.3.2 Large Time Solutions

For small times, the linear kinetic undercooling model holds for any value of β (although restrictions are imposed due to the linearisation of the relation $s_t(T_I)$; this is discussed in Sect. 5.2.4). The large time solution, on the other hand, takes different forms depending on whether β is greater, less than, or equal to 1. Davis [40] mentions three solution forms: $s \propto t^{1/2}$ for $\beta > 1$; $s \propto t$ for $\beta < 1$; and $s \propto t^{2/3}$ for $\beta = 1$. Here we will derive the three forms via similarity transformations and perturbation expansions.

Case $\beta < 1$

An obvious reduction of Eq. (5.13) comes through choosing $\alpha = 0$, which reduces the governing equations (5.15, 5.16) to

$$G'' = -\nu G' , \qquad 0 < \eta < \infty \tag{5.32}$$

$$G(0) = -s_t , \qquad G|_{\eta\to\infty} \to -1 , \qquad s_t = \nu , \qquad s(0) = 0 . \tag{5.33}$$

The case $\alpha = 0$ corresponds to $\eta = y = x - s$, which indicates a travelling wave solution. The travelling wave formulation differs from the other forms of similarity solution: the similarity transformation $\eta = y/t^\alpha$ with $\alpha \neq 0$ allows the far-field and initial conditions to be combined, that is $\eta \to \infty$ when $x \to \infty$ or $t \to 0$. With $\alpha = 0$, and so $\eta = y = x - s$ where $x > s$, the initial condition corresponds to $\eta = x$, while the far-field condition has $\eta \to \infty$, so the two do not coincide. Since initially the temperature is everywhere at $\theta(x, 0) = G(\eta) = -1$, the condition $G(0) = -s_t$ then requires $s_t = 1$.

The Stefan condition integrates immediately to give $s(t) = \nu t$, and this may be used to remove the time dependence in the boundary condition: $G(0) = -\nu$. Integrating the governing equation (5.32) and applying the boundary conditions in (5.33) leads to

$$G = -1 + (1 - \nu)e^{-\nu\eta} . \tag{5.34}$$

To determine the value of ν, we now use the above solution and the definition $\nu = -G'(0)/\beta$ to determine $\nu = 1 - \beta$ and hence

$$G = -1 + \beta e^{-\eta(1-\beta)} , \qquad s = (1 - \beta)t , \qquad s_t = (1 - \beta) . \tag{5.35}$$

Since the front propagates in the positive η direction, $s_t > 0$, this clarifies that the travelling wave solution is restricted to values $\beta < 1$.

The travelling wave solution describes a constant shape that propagates with a fixed velocity: consequently, the temperature and temperature gradient are constant at the moving front. From Eq. (5.35), we know that the velocity $s_t(0) = 1 - \beta$.

However, we noted above that for $T_I(t) \neq 0$, the initial velocity $s_t(0) = 1$: in other words, the solution of (5.35) does not satisfy the initial condition. The travelling wave solution is commonly quoted as an exact solution form but it often does not match an initial transient. If the initial condition is not satisfied but the travelling wave satisfies the governing equation and boundary conditions, then it may be considered a large time solution, only valid beyond the initial transient. With this in mind, we cannot apply $s(0) = 0$ and the results in the original variables are

$$\theta = -1 + \beta e^{-(1-\beta)(x-s)} , \qquad s = (1-\beta)t + C_0 , \qquad s_t = (1-\beta) . \tag{5.36}$$

The constant of integration C_0 must be determined either from matching to a small time solution or from some extra information. For example, in the studies of carbon capture of [118], the time taken for a CO_2 wave to escape the end of the adsorption column is used rather than an initial value.

Case $\beta > 1$

For $\beta > 1$, we can make progress by applying a large time substitution $t = \tau/\varepsilon$ to (5.15)–(5.16), but now take $\alpha = 1/2$ to remove the time dependence from the term $\alpha\eta t^{2\alpha-1}$. The Stefan condition (5.16b) then implies scaling $s = r/\varepsilon^{1/2}$, and the governing equations are

$$G'' = -\left(\nu + \frac{\eta}{2}\right) G' , \qquad 0 < \eta < \infty \tag{5.37}$$

$$G(0) = -\varepsilon^{1/2} r_\tau , \quad \tau^{1/2} r_\tau = \nu , \qquad G|_{\eta\to\infty} = -1 . \tag{5.38}$$

Note, in the previous example, we solved the boundary value problem for G and then found we could not apply the initial condition $s(0) = 0$, essentially forcing the assumption of a large time solution. Here we begin under the assumption of large time and so do not impose $s(0) = r(0) = 0$. The Stefan condition then immediately leads to $r = 2\nu\sqrt{\tau} + c_1$, where c_1 is an integration constant. The small parameter in the boundary condition suggests a series solution $G = G_0 + \mathcal{O}(\varepsilon^{1/2})$. The leading order problem is then identical to the system without kinetic undercooling, Eqs. (5.17, 5.18), but the initial condition no longer applies. Viewed from another angle, *we can interpret the no kinetic undercooling problem as a large time leading order approximation to the kinetic undercooling problem.* The first-order problem has a time-dependent boundary condition, and so, as for the small time solution of this section, only the leading order problem is consistent with the current similarity transformation.

Referring back to the solution of Eq. (5.21), we find

$$\theta \approx -1 + \frac{\operatorname{erfc}\,(x/(2\sqrt{t}) - C_1)}{\operatorname{erfc}\,(\nu)} , \qquad s \approx 2\nu\sqrt{t} + C_1 , \qquad s_t \approx \frac{\nu}{\sqrt{t}} . \tag{5.39}$$

The value of ν is found from the numerical solution of the transcendental equation (5.20). This has a unique solution for $\beta > 1$; no solution exists for $\beta \leq 1$ and consequently the above solution; Eqs. (5.39) are only valid for $\beta > 1$.

Case $\beta = 1$

For the case $\beta < 1$, the front velocity $s_t \approx 1-\beta$ is constant in time, and as $\beta \to 1^-$, the wave comes to a stop. For $\beta > 1$, the front velocity $s_t \approx \nu/\sqrt{t}$ decreases with increasing time. The value of ν increases with decreasing β such that as $\beta \to 1^+$, $\nu \to \infty$. In the intermediate case $\beta = 1$, we anticipate a different time dependence, which is not apparent from our previous analysis.

In contrast to the similarity transformation given in (5.12), we define

$$\eta = \frac{y}{t^\alpha} , \qquad F(y,t) = G(\eta, t) , \tag{5.40}$$

where G retains an explicit time dependence. The governing equations (5.11) become

$$\frac{\partial^2 G}{\partial \eta^2} = t^{2\alpha}\frac{\partial G}{\partial t} - \left(\alpha\eta t^{2\alpha-1} + t^\alpha s_t\right)\frac{\partial G}{\partial \eta} , \qquad 0 < \eta < \infty \tag{5.41}$$

$$G\big|_{\eta=0} = -s_t , \qquad G|_{\eta\to\infty} \to -1 , \qquad t^\alpha s_t = -\frac{\partial G}{\partial \eta}\bigg|_{\eta=0} ; \tag{5.42}$$

again we do not impose $s(0) = 0$.

We now apply a large time substitution to Eqs. (5.41, 5.42):

$$t = \frac{\tau}{\varepsilon} , \qquad s = \frac{r}{\varepsilon^\gamma} , \qquad G(\eta, t) = H(\eta, \tau) , \tag{5.43}$$

for the moment leaving the exponents α, γ unspecified. Then the Stefan condition in (5.42) becomes

$$\tau^\alpha r_\tau = \varepsilon^{\alpha+\gamma-1}\frac{\partial H}{\partial \eta}\bigg|_{\eta=0} .$$

We require $\alpha + \gamma - 1 = 0$ for the front to move (at leading order). Hence, the governing equations may be written as

$$\frac{\partial^2 H}{\partial \eta^2} = \tau^{2\alpha}\varepsilon^{1-2\alpha}\frac{\partial H}{\partial \tau} - \left(\alpha\eta\tau^{2\alpha-1}\varepsilon^{1-2\alpha} + \tau^\alpha r_\tau\right)\frac{\partial H}{\partial \eta} , \qquad 0 < \eta < \infty \tag{5.44}$$

$$H\big|_{\eta=0} = -\varepsilon^{\alpha} r_{\tau}\,, \qquad H\big|_{\eta\to\infty} \to -1\,, \qquad \tau^{\alpha} r_{\tau} = -\left.\frac{\partial H}{\partial \eta}\right|_{\eta=0}\,. \tag{5.45}$$

From examining (5.44) and (5.45a), it is clear that we require $0 < \alpha < 1/2$, in order that $\varepsilon^{1-2\alpha}$, $\varepsilon^{\alpha} \to 0$ as $\varepsilon \to 0$. The form of Eq. (5.44) suggests an expansion:

$$H(\eta,\tau) = H_0(\eta) + \varepsilon^{1-2\alpha}\tau^{\theta} H_1(\eta) + \ldots\,, \quad r(\tau) = r_0(\tau) + \varepsilon^{1-2\alpha} r_1(\tau) + \ldots\,, \tag{5.46}$$

where θ is left unspecified. Substituting these expansions into (5.44)–(5.45) gives the leading order problem:

$$H_0''(\eta) = -\nu_0 H_0'(\eta)\,, \qquad 0 < \eta < \infty \tag{5.47}$$

$$H_0(0) = 0\,, \qquad H_0\big|_{\eta\to\infty} \to -1\,, \qquad \tau^{\alpha} r_{0\tau} = \nu_0\,, \tag{5.48}$$

where $\nu_0 = -H_0'(0)$. The solution is

$$H_0(\eta) = -1 + e^{-\nu_0 \eta}\,, \qquad r_0 = \frac{\nu_0}{1-\alpha}\tau^{1-\alpha} + c_2\,. \tag{5.49}$$

Note, the definition of $\nu_0 = -H_0'(0)$ is automatically satisfied and provides no information. We must then look to the higher-order terms to determine ν_0. At first order, the boundary condition (5.45) for $H|_{\eta=0}$ leads to the balance $\varepsilon^{1-2\alpha}\tau^{\theta} H_1(0) = -\varepsilon^{\alpha} r_{0\tau}$. The terms balance if we choose $\alpha = 1/3$. Thus, the $\mathcal{O}(\varepsilon^{1/3})$ equations are

$$\tau^{\theta} H_1''(\eta) = -\frac{\eta}{3}\tau^{-1/3} H_0'(\eta) - \nu_0 \tau^{\theta} H_1'(\eta) - \tau^{1/3} r_{1\tau} H_0'(\eta)\,, \qquad 0 < \eta < \infty \tag{5.50}$$

$$\tau^{\theta} H_1(0) = -r_{0\tau}\,, \qquad H_1\big|_{\eta\to\infty} \to 0\,, \qquad \tau^{1/3} r_{1\tau} = -\tau^{\theta} H_1'(0)\,. \tag{5.51}$$

After replacing $r_{1\tau}$ in Eq. (5.50) with the final expression of (5.51), it becomes clear that to remove τ from the equations requires setting $\theta = -1/3$. Thus, (5.50), (5.51) reduce to

$$H_1''(\eta) = -\frac{\eta}{3} H_0'(\eta) - \nu_0 H_1'(\eta) - \nu_1 H_0'(\eta)\,, \qquad 0 < \eta < \infty \tag{5.52}$$

$$H_1(0) = -\nu_0\,, \qquad \tau^{2/3} r_{1\tau} = \nu_1\,, \qquad H_1\big|_{\eta\to\infty} \to 0\,, \tag{5.53}$$

where $\nu_1 = -H_1'(0)$. Since H_0 is known, the solution to this system is simple to calculate

$$H_1 = -\frac{1}{6\nu_0^2}\left[2 + 6\nu_0^2\nu_1\eta + 2\nu_0\eta + \nu_0^2\eta^2\right]e^{-\nu_0\eta}\,. \tag{5.54}$$

Substituting for $H_1(0)$ into Eq. (5.53a) determines ν_0

$$\nu_0 = \left(\frac{1}{3}\right)^{1/3} . \tag{5.55}$$

Finally, substituting for α and ν_0, we may write down the position of the moving front:

$$r_0 = \left(\frac{9}{8}\right)^{1/3} \tau^{2/3} + c_2 . \tag{5.56}$$

Note, the dependence $r \sim \tau^{2/3}$ is quoted in [33, 76, 84] (but omitting the multiplicative constant). Due to the complicated analysis to arrive at this dependence, it is not surprising that this is the one β value that is not proved in [40]. Finally, in the original notation, we have

$$\theta \approx -1 + \exp\left(\frac{x-s}{(3t)^{1/3}}\right) , \qquad s \approx \left(\frac{9}{8}\right)^{1/3} t^{2/3} + C_2 , \qquad s_t \approx (3t)^{-1/3} . \tag{5.57}$$

In Fig. 5.2, we show two sets of results taken from [52]. In this paper, the authors carried out the above analysis as well as applying the TIM with $n = 3.57$ to determine approximate solution forms. The TIM approach reduces the system to a single second-order ordinary differential equation for s. It is solved numerically using a MATLAB function. Results were verified by applying a numerical scheme to Eqs. (5.8), following the method described in [92]. The three curves in the figures represent the numerical solution (solid line), the TIM solution (dashed line),

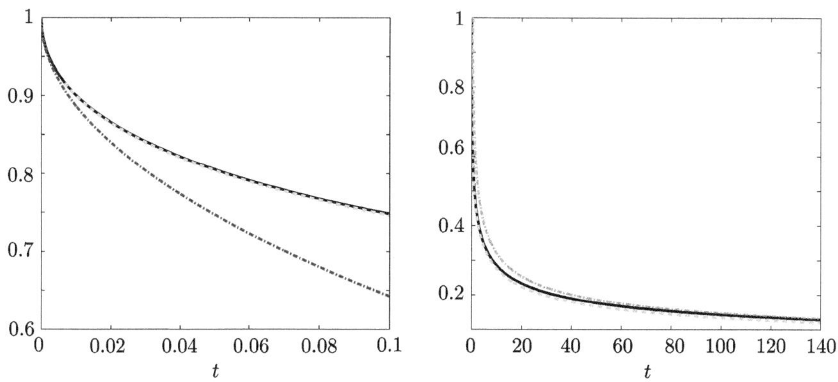

Fig. 5.2 Velocity of the interface with linear kinetic undercooling and $\beta = 1$ with (left) $t \in [0, 0.1]$, (right) $t \in [0, 140]$. The sets of curves denote the numerical solution (solid line) of Eqs. (5.8), a TIM solution (green, dashed), and left figure, the small time solution of Eq. (5.31) (red, dash-dot line), and right figure, large time solution of Eq. (5.57) (purple, dash-dotted). Parameter values for copper have been used; see Table 5.1

and on the left diagram, the small time solution (dash-dot) given by Eq. (5.31), and on the right diagram, the large time solution of Eq. (5.57). The small time asymptotic solution clearly loses accuracy very rapidly. We note that this solution does have a time restriction, since the leading order term in the expression for s must be significantly greater than the first order; see Eq. (5.31); this suggests $t \ll 9\pi/(16\beta^2)$. The TIM provides an excellent approximation for all time. The small time solution loses accuracy as time increases; on the right-hand plot, we see that the large time solution improves as t increases. The TIM again provides good agreement with the numerical solution, but for t approximately greater than 80, the large time solution becomes more accurate than the TIM.

5.2.4 Nonlinear Kinetic Undercooling

Theoretically, the Neumann solution only holds for values $\beta > 1$. By employing the linear undercooling approximation, the solution domain could be extended to include $\beta \leq 1$. If we take parameter values for salol L_m, c_l, given in Table 5.1, we find $\beta \to 1$ as $\Delta T \to 57\,\text{K}$. This is where the Neumann solution predicts the velocity tending to infinity. Examining Fig. 5.1 (right), it is clear that at this level of undercooling, the velocity is significantly below its maximum value of 4.5×10^{-5} m/s. Further, the linear model only appears to be valid for $\Delta T < 10\,\text{K}$, which corresponds to $\beta > 5.7$. For copper, a similar argument indicates the linear model holds for $\beta > 4$. For values of β greater than 5.7 or 4, with salol or copper, neither the Neumann nor linear approximations will provide physically realistic solutions. Consequently, to go beyond these values, we must apply the full nonlinear relation (5.9).

The similarity solutions of the previous sections were tractable because of the simple dependencies at the boundary $x = s(t)$. The nonlinear relation (5.9) prevents these simple reductions, and so in this section, we limit our analysis to small time and travelling wave solutions.

5.2.4.1 Small Time Solutions

Following the analysis of Sect. 5.2.3, we again obtain the system (5.23)–(5.24) with the one change being that the right-hand side of the first boundary condition in (5.24a) is replaced with T_I. Then, using this condition to replace T_I in the interface temperature equation (5.9), we find

$$s_t = \left[1 - t^{1/2}G(0)\right] \exp\left\{\frac{Q\left[-1 + t^{1/2}G(0)\right]}{P + \left[-1 + t^{1/2}G(0)\right]}\right\}. \tag{5.58}$$

Taking the small time substitution $t = \varepsilon\tau$, $s = \varepsilon r$, and expanding the exponential term using a Taylor series expansion brings us to the leading order problem specified by Eqs. (5.27, 5.28) but with the condition $r_{0\tau} = 1$ replaced by

$$r_{0\tau} = \exp\left(-\frac{Q}{P-1}\right). \tag{5.59}$$

The leading order solution to this system corresponds to Eq. (5.29):

$$G_0 = \beta \exp\left(-\frac{Q}{P-1}\right)\left[\frac{2}{\sqrt{\pi}}e^{-\eta^2/4} - \eta \operatorname{erfc}\left(\frac{\eta}{2}\right)\right], \tag{5.60}$$

$$r_0 = \tau \exp\left(-\frac{Q}{P-1}\right). \tag{5.61}$$

In keeping with the previous analysis, we cannot progress to solving for G_1 but can determine r_1,

$$r_1 = -\exp\left(-\frac{2Q}{P-1}\right)\left[1 - \frac{QP}{(P-1)^2}\right]\tau^{3/2}\frac{4\beta}{3\sqrt{\pi}}. \tag{5.62}$$

Returning to the original variables gives

$$\theta \approx -1 + t^{1/2}\beta \exp\left(-\frac{Q}{P-1}\right)\left[\frac{2}{\sqrt{\pi}}\exp\left(-\frac{(x-s)^2}{4t}\right) - \left(\frac{x-s}{t^{1/2}}\right)\operatorname{erfc}\left(\frac{x-s}{2t^{1/2}}\right)\right], \tag{5.63a}$$

$$s \approx \exp\left(-\frac{Q}{P-1}\right)\left\{1 - \exp\left(-\frac{Q}{P-1}\right)\left[1 - \frac{QP}{(P-1)^2}\right]\frac{4\beta}{3\sqrt{\pi}}t^{1/2}\right\} t, \tag{5.63b}$$

$$s_t \approx \exp\left(-\frac{Q}{P-1}\right)\left\{1 - \exp\left(-\frac{Q}{P-1}\right)\left[1 - \frac{QP}{(P-1)^2}\right]\frac{2\beta}{\sqrt{\pi}}t^{1/2}\right\}. \tag{5.63c}$$

These are valid for any $\beta > 0$.

5.2.4.2 Large Time Solutions

Case $\beta < 1$

As stated, we will restrict our attention to the travelling wave reduction, which, as we will see, requires $\beta < 1$.

The nonlinear undercooling assumption has little effect on the travelling wave analysis of Sect. 5.2.3, resulting in

$$G = -1 + (1 + T_I)e^{-\nu\eta}\,, \tag{5.64}$$

where $T_I = \beta - 1$. The wave speed, ν, is determined by substituting $s_t = \nu$ into Eq. (5.9)

$$\nu = s_t = (1 - \beta)\exp\left(\frac{Q(\beta - 1)}{P + \beta - 1}\right). \tag{5.65}$$

The temperature, position of the front, and velocity in the original variables are

$$\theta = -1 + \beta\exp(-\nu(x - s))\,, \tag{5.66}$$

$$s = (1 - \beta)\exp\left(\frac{Q(\beta - 1)}{P + \beta - 1}\right)t + C_0\,, \quad s_t = (1 - \beta)\exp\left(\frac{Q(\beta - 1)}{P + \beta - 1}\right). \tag{5.67}$$

As in the linear case, to maintain a positive velocity, the solution is restricted to values of $\beta < 1$ and, since it cannot satisfy the initial condition, should be considered a large time approximation.

In Fig. 5.3, we present the evolution of the wave front for small (left) and large (right) times with $\beta = 1$. On the left figure, we show the numerical solution (solid line) of Eqs. (5.8), a TIM solution (dashed) with $n = 3.57$, and the small time solution of Eq. (5.63c) (dash-dot line) for $t \in [0, 0.1]$. The TIM solution,

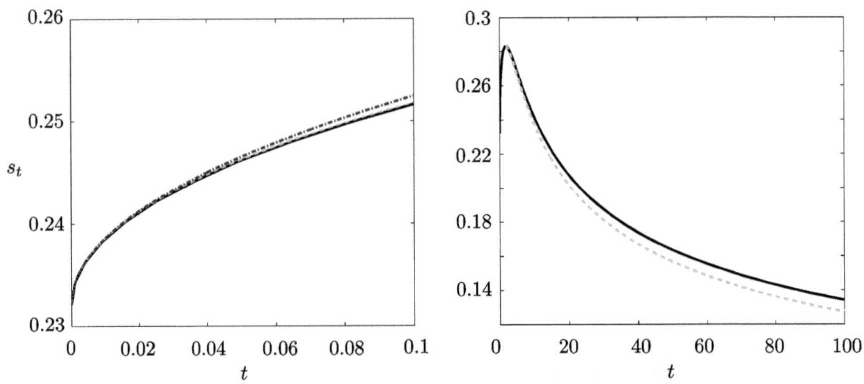

Fig. 5.3 Velocity of the interface with nonlinear kinetic undercooling and $\beta = 1$. The sets of curves denote, left figure, the numerical solution (solid line) of Eqs. (5.8), a TIM solution (green, dashed), and the small time solution of Eq. (5.63c) (red, dash-dot line) for $t \in [0, 0.1]$, and right figure, numerical solution (solid line) and TIM solution (green, dashed). Parameter values correspond to copper; see Table 5.1, such that $Q = 3.5811$, $P = 3.453$

developed in [52], reduces to two coupled nonlinear first-order ordinary differential equations, which are solved numerically. For very small times, the two approximate solutions are very accurate but the asymptotic solution slowly loses accuracy as time increases. As before, this is bounded by a time restriction. Decreasing β would improve the accuracy of the asymptotic solution. An important difference between this solution and the linear (and zero) undercooling cases is that the initial speed now increases with time. This is a consequence of the form shown in Fig. 5.1; in this example, the initial undercooling is such that we start to the right of the peak on the $s_t(\Delta T)$ curve. As time proceeds and T_I approaches T_m, we will move to the left and so observe an initial increase in s_t followed by a decrease as we pass the peak. This behaviour is apparent in the right-hand plot of Fig. 5.3, which shows a numerical and TIM solution for $t \in [0, 100]$ (we have no asymptotic solution here). For $t \in [0, 5]$ (approximately) s_t increases to a maximum of just above 0.28 and then slowly decreases with the $t^{-1/3}$ behaviour predicted earlier.

5.3 Nanocrystal Growth

We now move to a more modelling-oriented problem and investigate the growth of nanocrystals from a colloidal solution. Nanocrystals are a particular form of nanoparticle, where the particle has a crystalline structure. Precipitation of nanocrystals from solution is one of the most widely used synthesis methods. The typical strategy is to cause a short nucleation burst that creates a large number of nuclei in a short space of time. These form the seeds for the later particle growth stage. The purpose of the following analysis is to understand, and so better control, the process of nanocrystal growth.

Before stating the mathematical model, we introduce the Ostwald-Freundlich condition (which relates the radius of the particle, r_p, to its solubility, S) and also the concept of Ostwald ripening. The Ostwald-Freundlich condition can be written as

$$S = S_\infty \exp\left(\frac{2\sigma V_M}{r_p R_G T}\right) \equiv S_\infty \exp\left(\frac{\alpha_c}{r_p}\right), \tag{5.68}$$

where S_∞ is the solubility of the bulk material, σ the interfacial energy, V_M the molar volume, R_G the universal gas constant, and T the temperature. The capillary length $\alpha_c = 2\sigma V_M/(R_G T)$ defines the length-scale below which curvature-induced solubility is significant. Equation (5.68) shows that the particle solubility increases as the size decreases. If the solubility of the crystal is below the concentration of the bulk solution $S < c_b$, then monomer molecules diffuse from the bulk towards the particle to react with the surface and the particle grows, whereas if $S > c_b$, the molecules detach and the particle shrinks. The Ostwald-Freundlich condition then provides a way to estimate whether a nanocrystal will shrink or grow. In the limit of small α_c/r_p, it may be written as $S \approx S_\infty(1+\alpha_c/r_p)$, demonstrating an analogy with the Gibbs-Thomson equation (5.4).

Ostwald ripening is the phenomenon whereby small crystals dissolve and the material may then redeposit onto larger crystals. In the manufacture of crystals from a colloidal solution, it can lead to a spread in the size distribution (defocussing) where the practical goal is invariably to produce a narrow range of crystal sizes. With stable crystals, Ostwald ripening can be an extremely slow process. A famous experiment of Faraday (in the basement of the Royal Institution in London) on colloidal gold has demonstrated Ostwald ripening over a period of 150 years. Ostwald ripening may then be viewed as a direct consequence of the Ostwald-Freundlich condition (5.68).

In the following, we take the approach of Myers and Fanelli [47, 105] and begin by analysing the growth of a single particle, which is the basic building block for a model involving many particles. We focus on a spherical nanoparticle, with radius $r_p(t)$ in a system of particles. It is assumed that the particles are separated by large but finite distances compared to their radius. They remain approximately spherical throughout the process and so the mathematical problem is radially symmetric. Following the standard La Mer model [71], the growth process begins with a short nucleation burst. This may be caused, for example, by supersaturating a solution under pressure and then releasing the pressure. The mathematical model starts immediately after the nucleation phase.

The crystal growth is described by a standard diffusion equation in a spherical co-ordinate system:

$$\frac{\partial c}{\partial t} = \frac{D}{r^2}\frac{\partial}{\partial r}\left(r^2\frac{\partial c}{\partial r}\right), \qquad r > r_p(t) \tag{5.69}$$

$$\frac{dr_p}{dt} = V_m D \left.\frac{\partial c}{\partial r}\right|_{r_p}, \tag{5.70}$$

where c is the monomer concentration, D the diffusion coefficient and V_m the molar volume. The system is closed by the following boundary and initial conditions:

$$c|_{r_p+\delta} = c_b(t), \quad c(r_p, t) = c_i(t), \quad c(r, 0) = c_b(0), \quad r_p(0) = r_{p0}. \tag{5.71}$$

The concentration $c_b(t)$ is the bulk or far-field concentration of monomer in the solution that decreases with time as monomer is taken up by the nanocrystals. Instead of applying this condition as $r \to \infty$, we have imposed a boundary layer thickness $\delta(t)$, despite being completely aware that the above diffusion equation has infinite speed of propagation and so, strictly $\delta(t) = \infty$ for all time $t > 0$. (This unphysical, but widely accepted, speed is one motivation for the models presented in Chap. 6). There are two reasons for the introduction of $\delta(t)$. Firstly, it is standard in the chemistry literature to define a diffusion boundary layer and, when dealing with multidisciplinary topics, it is important, whenever possible, to relate the model to existing work. Secondly, by including $\delta(t)$, we are able to explain an issue that has pervaded the literature. Later we will assume $\delta \gg r_p$, so its inclusion should not hurt any mathematician's feelings. The initial condition for the crystal radius is $r_p(0) = r_{p0} \neq 0$ since the growth comes from finite seed crystals caused by the burst nucleation process. The quantity $c_i(t)$ represents the concentration at the crystal boundary. This is a difficult quantity to measure, so below we will relate it to the mass flux and solubility.

The above model is clearly similar to that presented in Sect. 2.1.4, but in contrast to that model, here the concentration in the far field and at the crystal boundary varies with time; also the crystal starts off with a finite radius. These factors all prevent the use of the similarity solution employed in the previous analysis.

The bulk concentration is determined through mass conservation of the monomer in the particle and in the surrounding solution:

$$\frac{1}{N_0} M_p c_0 = M_p c_b(t) \left[\frac{1}{N_0} - \frac{4\pi}{3} {r_p}^3 \right] + \frac{4\pi}{3} \rho_p {r_p}^3, \tag{5.72}$$

where c_0 is the initial monomer concentration (measured before the seed crystals appear), ρ_p is the particle density, M_p is the molar mass, and N_0 is the particle population density (the number of particles per unit volume). The left-hand side represents the mass of monomer in the volume occupied by a single crystal. The right-hand side contains two components: the first is the mass of monomer in the solution surrounding the crystal (this volume consists of the original region minus the volume occupied by the growing crystal). The second component is the mass of the crystal. If the system is dilute, then the crystal occupies a small fraction of the liquid region, so $4\pi N_0 {r_p}^3/3 \ll 1$. Writing the molar volume $V_M = M_p/\rho_p$, Eq. (5.72) reduces to

$$c_b(t) \approx c_0 - \frac{4\pi N_0}{3V_M} {r_p}^3, \tag{5.73}$$

valid for a dilute system.

The concentration adjacent to the crystal surface is an unknown function of time $c_i(t)$. To determine an expression for c_i, we consider two equivalent relations for the mass flux at the particle surface, J. Fick's first law states that the flux of monomer passing through a spherical surface of radius r is

$$J = 4\pi r^2 D \frac{\partial c}{\partial r}. \tag{5.74}$$

This relation is key to deriving the "Stefan" condition, Eq. (5.70), which comes from the mass balance:

$$\rho_p \frac{d}{dt}\left(\frac{4}{3}\pi r_p^3 \right) = M_p 4\pi r^2 D \frac{\partial c}{\partial r}. \tag{5.75}$$

At the surface of the sphere, some unspecified reaction occurs causing monomer to attach. The simplest realistic model for this process is a first-order reaction equation:

$$J = 4\pi r^2 k\,(c_i - S)|_{r=r_p}\,, \tag{5.76}$$

where k is the reaction rate. This quantifies the reaction and is assumed to be constant. Equating (5.74) and (5.76) gives

$$c_i = S + \left.\frac{D}{k}\frac{\partial c}{\partial r}\right|_{r=r_p}\,, \tag{5.77}$$

which defines the concentration c_i for the surface condition of (5.71), $c(r_p, t) = c_i(t)$, and recall $S = S_\infty \exp(\alpha_c/r_p)$. Clearly, the above system is analogous to a one-phase Stefan problem albeit with rather unconventional boundary conditions.

The system is nondimensionalised via

$$\hat{r} = \frac{r}{r_{p,0}}\,, \quad \hat{r}_p = \frac{r_p}{r_{p,0}}\,, \quad \hat{t} = \frac{t}{\tau}\,, \quad \hat{c} = \frac{c - S_0}{\Delta c}\,, \quad \hat{S} = \frac{S - S_0}{\Delta c}\,, \tag{5.78}$$

where $\Delta c = c_b(0) - S_0$ represents the driving force for particle growth and $S_0 = S(0) = S_\infty \exp(\alpha_c/r_{p,0})$ is the initial particle solubility. The concentration and growth equations yield two possible time-scales, the diffusion scale $\tau_D = r_{p,0}^2/D$ and the reaction scale $\tau_R = r_{p,0}^2/(V_M D \Delta c)$, respectively. To focus on particle growth, we choose the reaction time-scale τ_R, and the system is now transformed to

$$\varepsilon\frac{\partial c}{\partial t} = \frac{1}{r^2}\frac{\partial}{\partial r}\left(r^2\frac{\partial c}{\partial r}\right)\,, \qquad \frac{dr_p}{dt} = \left.\frac{\partial c}{\partial r}\right|_{r=r_p}\,, \tag{5.79a}$$

$$c(r_p, t) = S + Da\left.\frac{\partial c}{\partial r}\right|_{r=r_p}\,, \qquad c(r_p + \delta, t) = c_b(t) = c_0 - \gamma r_p^3\,, \tag{5.79b}$$

$$S = \frac{S_\infty}{\Delta c}\left(e^{\omega/r_p} - e^{\omega}\right)\,, \qquad c(r, 0) = 1\,, \qquad r_p(0) = 1\,, \tag{5.79c}$$

where

$$\varepsilon = V_M \Delta c\,, \qquad Da = \frac{D}{k r_{p,0}}\,, \qquad \omega = \frac{\alpha_c}{r_{p,0}}\,, \qquad \gamma = \frac{4\pi N_0 r_{p,0}^3}{3 V_M \Delta c}\,, \tag{5.79d}$$

and the nondimensional c_0 is related to the dimensional counterpart by $(c_0 - S_0)/\Delta c$. The above system contains a number of nondimensional groups. The first, ε, is generally very small for nanoparticle growth. For example, Peng et al. [125] studied cadmium selenide (CdSe) nanoparticles, with a capillary length of 6 nm and initial radii in the range 1−100 nm, with parameter values such that $\varepsilon = \mathcal{O}(10^{-3})$; see Table 5.2. In general, for crystal growth, it should be expected that $\varepsilon \ll 1$. The parameter Da is a form of inverse Damköhler number, measuring the relative magnitude of diffusion to surface reactions.

The presence of the small parameter in the concentration equation suggests a pseudo-steady solution (and the small parameter is truly small; nanocrystal growth is significantly slower than diffusion). Integrating and applying the boundary conditions gives

$$c = -\frac{A}{r} + B\,, \tag{5.80}$$

where

$$A = \frac{r_p^2(r_p + \delta)(c_b - S)}{r_p\delta + Da(r_p + \delta)}\,, \qquad B = S + A\left(\frac{1}{r_p} + \frac{Da}{r_p^2}\right). \tag{5.81}$$

There is no way to calculate $\delta(t)$ in the pseudo-steady approach; a time-dependent treatment is required (an excellent opportunity to apply the TIM described in Sect. 2.2.4.2 or a perturbation expansion to find higher-order terms). However, the issue may be avoided by assuming that, except for during an initial transient, $r_p \ll \delta$. This reduces the concentration to

$$c = c_b - \frac{r_p^2(c_b - S)}{(r_p + Da)r}\,. \tag{5.82}$$

Substituting (5.80) into the growth condition (5.79a) leads to

$$\frac{dr_p}{dt} = \frac{c_b(r_p) - S(r_p)}{Da + r_p} = \frac{c_0 - \gamma r_p^3 - S(r_p)}{Da + r_p}\,. \tag{5.83}$$

Provided the particle radius is significantly greater than the capillary length, $r_p \gg \omega$, we may set $S(r_p) = S_{eq}$ to be constant, and then it is a simple matter to integrate equation (5.83). The result of the integration is an implicit equation for $t(r_p)$ involving arctan and logarithmic terms. In [105], it is demonstrated that the arctan terms are at most three orders of magnitude smaller than the logarithmic terms, so their relative importance is commensurate with previously neglected terms. In which case, the relation may be inverted to give

$$r_p = \frac{r_m}{2}\frac{\left[1 + 2f(r_{ps})\exp\left(\frac{t-t_s}{G}\right) - \sqrt{-3 + 12f(r_{ps})\exp\left(\frac{t-t_s}{G}\right)}\right]}{\left[-1 + f(r_{ps})\exp\left(\frac{t-t_s}{G}\right)\right]}\,, \tag{5.84}$$

where r_m is the (nondimensional) experimental maximum radius. This is introduced since the assumption that $\delta \gg 1$ means the model cannot be applied at $t = 0$;

consequently, we use the final time radius to calculate the constant of integration. The time t_s and associated radius $r_p(t_s) = r_{ps}$ will be clarified below; the function $f(r_{ps}) = (r_m^2 + r_m r_{ps} + r_{ps}^2)/(r_m - r_{ps})^2$. This model contains a single unknown, the parameter G defined by

$$G = \frac{1}{6ab}\frac{ak + bD}{akbD}, \qquad a^3 = V_m(c_0 - S_{eq}), \qquad b^3 = \frac{4}{3}\pi N_0, \tag{5.85}$$

where we define the equilibrium solubility $S_{eq} = S_\infty \exp(\omega/r_m)/\Delta c$. In fact, G combines two unknowns from the mathematical model, the kinetic coefficient k and the monomer diffusion coefficient D. It is determined by comparison with experimental data. Equation (5.84) was the first explicit solution to Eq. (5.83) presented in the literature. Previous solutions were based on assumptions concerning the relative size of diffusion to reaction rate, so removing either k or D from the calculation. This led to two schools of thought, one contending that diffusion drove the crystal growth and the other that it was the reaction. Since k, D only appear in the solution through G, where ak, bD are interchangeable, it is argued in [105] that the two effects are indistinguishable in a pseudo-steady model. This was confirmed by comparison with a variety of published results where the calculated values of ak, bD were almost identical, within the level of approximation applied (e.g. in [149] the values $ak \approx 3.2 \times 10^{-10}$ while $bD \approx 3 \times 10^{-10}$).

A second important conclusion from [105] is that there actually exist two distinct growth regimes (perhaps this is obvious from a mathematical viewpoint and indeed there may be more than two). It is argued that since initially the boundary layer thickness, $\delta(0) = 0$, is smaller than the particle radius, the assumption that $r_p \ll \delta$ is not valid during the early growth. In fact, this is apparent from experimental data, which shows an initial rapid growth rate not consistent with the solution of (5.83). Consequently, when calculating G, the very first data points should be neglected: the time t_s and radius r_{ps} are then chosen as the values from the first data point used in the calculations. The choice of r_p, t_s is not unique. Provided t_s is chosen within the later growth stage, the choice should not greatly affect the results. However, in general, we would prefer to choose t_s as close to the switch between the initial and later stages as possible to maximise the number of data points for the fitting. This is discussed in further detail in [105].

In Fig. 5.4, we present two sets of results for the growth of cadmium selenide nanocrystals. Appropriate data for the model, taken from the experimental study of Peng et al., [125] is provided in Table 5.2. Note, not all parameters are stated in Peng's paper; the values D, k come from the fitting to obtain G in the model, while the solubility and population density are inferred from other data. The left-hand figure shows values for the crystal radius data (circles) taken from [125] starting from a seed of around 2 nm and growing to approximately 3.8 nm after 2.5 hours. The first three points illustrate the issue with the two distinct growth phases in that they follow a steep, almost straight, line; subsequent data points follow the pseudo-steady model. The solid line comes from a least-squares fit of the data points to (5.84), starting from the fourth point. This fit determines $G = 958$. The right-hand

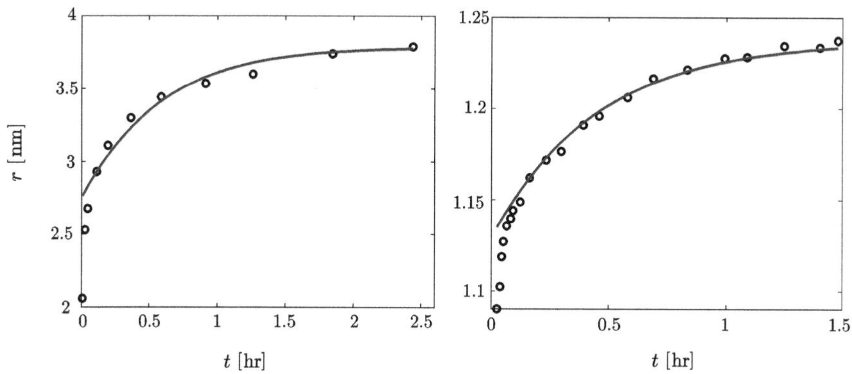

Fig. 5.4 Comparison of pseudo-steady model (solid line) with experimental data (circles) for the growth of cadmium selenide nanocrystals. Left data taken from [125] and right data from [149]

Table 5.2 Physical parameters for the cadmium selenide (CdSe) nanoparticle synthesis method used by Peng et al. [125]. The parameters in italics are not given explicitly and are obtained via a fitting approach or inferred from other data

Quantity	Symbol	Value	Units
Universal gas constant	R_G	8.31	J/(mol K)
Density	ρ_p	5816	kg/m^{-3}
Molar volume	V_M	3.29×10^{-5}	m^3/mol
Molar mass	M_p	0.19	kg/mol
Solution temperature	T	573.15	K
Surface energy	σ	0.44	J/m^2
Capillary length	α_c	6.00×10^{-9}	m
Initial bulk concentration	c_0	55.33	mol/m^3
Volume of the liquid	V	7.21×10^{-6}	m^3
Maximum particle radius	r_m	3.78×10^{-9}	m
Diffusion coefficient	D	3.01×10^{-18}	m^2/s
Reaction rate	k	7.97×10^{-10}	m/s
Solubility of bulk material	S_∞	5.53×10^{-2}	mol/m^3
Population density	N_0	8.04×10^{21}	No/m^3

plot involves data taken from a study by Su et al. [149] for the growth of CdSe nanocrystals. The seed is approximately 1.08 nm and the radius reaches 1.24 nm after 1.5 hours. The least-squares fit (here taken from the ninth data point) indicates $G = 789$. The first four points again show a steep linear growth, followed by a transition region between the linear phase and pseudo-steady growth. In Fanelli et al [47], a variety of previous models for the evolution of the crystal radius are presented. These models contained between two and eight fitting parameters. Equation (5.84), with a single unknown, was shown to provide better agreement with the data than all the previous models.

In Fanelli and Myers [47], the single particle model of [105] is extended to deal with N particles where N is arbitrarily large and may decrease with time. The possible decrease is due to Ostwald ripening, which cannot be predicted by a single particle model. To investigate this model, we now define the particle radii, initial radii, and solubilities as r_i, $r_{i,0}$, and S_i, respectively, where $i = 1, \ldots, N$ represents the ith crystal. The only difference in the nondimensionalisation compared to the single crystal version being that the length-scale $r_{p,0}$ is replaced by the mean value $\bar{r}_{i,0} = \Sigma_{i=1..N}(r_{i,0}/N)$, where the overbar is employed to denote the mean. This affects the concentration scale through the initial solubility, which now becomes $\overline{S}_{i,0} = S_\infty e^{\alpha_c/\bar{r}_{i,0}}$; then $\overline{\Delta c} = c_{b,0} - \overline{S}_{i,0}$. Hence in what follows, all dimensionless parameters defined in Eq. (5.79d) carry an overbar to reflect the change in length-scale.

Under the pseudo-steady approximation, and assuming that there are no inter-particle diffusional interactions, the growth of each particle is now described by an equation of the form (5.83) (but involving average quantities). This requires an expression for the bulk concentration that must account for the fact that all particles remove monomer and that the volume is now N times that of the single particle:

$$c_0 = c_b(t)\left[1 - \frac{V_M\overline{\Delta c}}{N}\sum_{i=1}^{N} {r_i}^3\right] + \frac{\overline{\gamma}}{N}\sum_{i=1}^{N} {r_i}^3. \tag{5.86}$$

Again assuming that the solution is sufficiently dilute that $(V_M\overline{\Delta c}/N)\sum_{i=1}^{N} r_i^3 \ll 1$, we obtain

$$c_b(t) \approx c_0 - \frac{\overline{\gamma}}{N}\sum_{i=1}^{N} r_i^3. \tag{5.87}$$

In the limit where all particles are identical $\sum_{i=1..N} r_i^3 = N r_i^3$, the single particle model of Eq. (5.73) is retrieved. The problem is now governed by the system of differential equations

$$\frac{dr_i}{dt} = \frac{c_b(r_i) - \overline{S}(r_i)}{\overline{Da} + r_i} = \frac{c_0 - (\overline{\gamma}/N)\sum_{i=1}^{N} r_i^3 - \overline{S}(r_i)}{\overline{Da} + r_i}, \tag{5.88}$$

for each $i = 1, \ldots, N$ and

$$\overline{Da} = \frac{D}{k\bar{r}_{i,0}}, \qquad \overline{\gamma} = \frac{4\pi N_0 \bar{r}_{i,0}^3}{3V_M\overline{\Delta c}}, \qquad \overline{\omega} = \frac{\alpha_c}{\bar{r}_{i,0}}. \tag{5.89}$$

Equation (5.88) provides a system of N nonlinear ordinary differential equations that must be solved numerically to determine r_i.

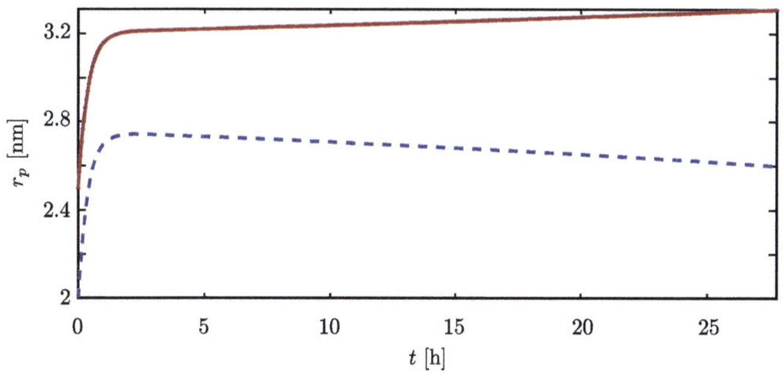

Fig. 5.5 Dimensional evolution of the radius of two cadmium selenide nanoparticles, with initial radii of 2 nm (dashed blue line) and 2.5 nm (solid red line). The solution is obtained through solving equation (5.88), with $N = 2$, numerically

To simulate the experiments of Peng et al. [125] in [47], a distribution of N nanoparticles was considered where the initial distribution was generated randomly (using the MATLAB random number generator), with an initial mean radius $\bar{r}_{i,0}$ of 2.92 nm and a standard deviation of $\sigma_o = 8.9\%$. In the numerical solution, if a particle decreased below the seed crystal size of 2 nm, it was assumed to break up, and all the monomer returns to the bulk concentration. Results showed that the mean radius increased with time in a manner similar to that of the single particle model and that as the number of particles in the simulation increased, the result approached that predicted by equation (5.84): for $N = 1000$, the maximum difference was less than 0.15%. It was concluded that the analytical approximation of equation (5.84) was sufficient for predicting the averaged evolution of a large number of nanoparticles. Taking the argument further, the model with $N = 2$ was sufficient to describe a bimodal distribution of a large number of particles.

In Fig. 5.5, we show the results of simulations for (5.88) with $N = 2$ and initial radii 2 and 2.5 nm, or equivalently two distributions of particles with a mean radii of 2 and 2.5 nm. The larger particle initially grows rapidly but at around 2 hours the growth rate slows and becomes approximately linear. The smaller particle suffers a very different fate, initial rapid growth followed by a period of slowly dissolving. This may be explained by considering Fig. 5.6. The dotted line represents the bulk concentration, which is initially much greater than the solubility of both particles. As the particles grow using up monomer, the bulk concentration rapidly decreases. With a single particle, or a group of equal size particles, when the bulk concentration approaches the particle solubility, the growth stops since $dr_p/dt \propto (c_b - S)$. With two particle sizes, the solubility of the smaller may approach the bulk concentration, so slowing its growth, but the larger continues to grow and use up monomer, thus taking the bulk concentration below the solubility of the other. On the figure, we may observe that at around 2 hours, the bulk concentration drops below the solubility of the smaller particle. For this reason, the second particle starts to lose mass and the Ostwald ripening process begins. Since we track the evolution over more than 25 hours, it is clear that for CdSe, this is a slow process.

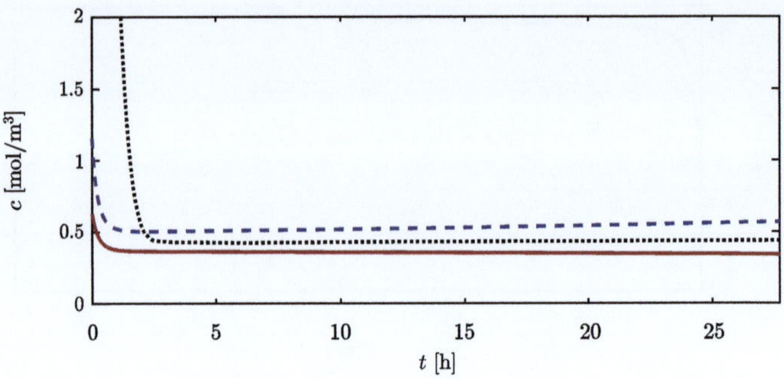

Fig. 5.6 Dimensional evolution of the bulk concentration (5.87) (dotted line) and solubility (5.79c) of two cadmium selenide nanoparticles with initial radii of 2 nm (dashed line) and 2.5 nm (solid line). When the solubility of the smaller particle exceeds the bulk value, at around 2 hours, the smaller particle starts to decrease in size

The main goal of the work described in [47, 105] was to understand and ultimately improve the growth process. For commercial applications, crystal growth is a multistage process, with extra monomer being added whenever it is judged that the concentration is too low [12]. Generally, this is an ad hoc process, with no clear rules on when or how much to add. One way the model presented here can aid in this process is by allowing different injection strategies to be tested in a simple way.

Firstly, we note that Eq. (5.83) indicates a maximum growth rate when $c_b - S$ is maximum. This suggests maintaining a high bulk concentration. However, if too much monomer is added, the concentration will surpass the saturation level, leading to the nucleation of more crystals and an undesirable, unfocussed size distribution. With the model presented above, the average radius is known, hence so is the bulk concentration, via the mass balance (5.73). Hence, at any given time, we know how much monomer is in solution and consequently how much may be added to keep below the spontaneous nucleation threshold. In theory, a continuous drip, which matches the rate at which monomer is lost, could be used to replace monomer and maintain the concentration at a constant high level. In practice, monomer is usually added in distinct stages; the above solution can determine the maximum possible amount that may be added without causing spontaneous generation.

A multi-injection solution requires new initial conditions at each injection stage. This means, for example, that the initial radius for the second stage (when new monomer is first injected) would be the value determined at the end of the first stage. The value of a in Eq. (5.85) must also be adjusted since the new initial concentration will be whatever remains at the end of the first stage plus the amount injected. This value then affects G.

In Fig. 5.7, the evolution of the radius is shown for an experiment with similar conditions to those studied in [125]. Hence, all parameter values from Table 5.2 are kept the same, except the initial value $c_0 = 50$ mol/m^3 and subsequently 50 and 100 mol/m^3 are added at different times. The solid curves have the addition at 0.25 and 0.5 hours and the dashed curves after 1 and 3 hours. In both cases, the final

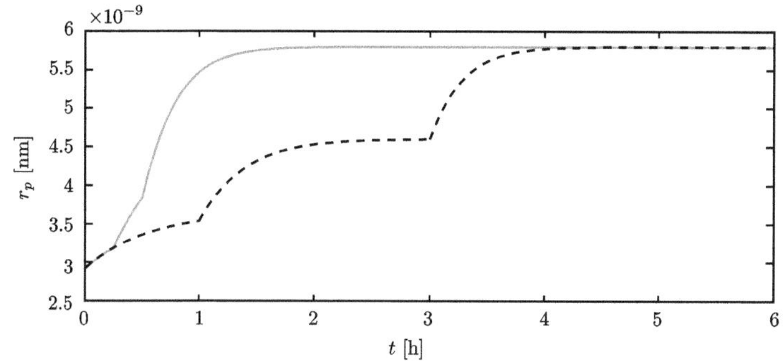

Fig. 5.7 Dimensional evolution of the radius of cadmium selenide nanoparticles with additional monomer injected at 0.25 and 0.5 hours (solid line) and 1 and 3 hours (dashed line)

radius is around 5.8 nm, but in the first case, this is achieved after 2 hours and in the second close to 5 hours. This demonstrates the ability of the method to guide injection strategies and optimise the growth process.

5.4 Nanoparticle Melting

Nanoparticle melting has been studied theoretically by a number of research groups around the world, slowly pushing forwards our understanding of how phase change occurs at the smallest dimensions; see [9, 10, 48, 51, 53, 82, 132, 157].

In an investigation of the effects of density variation on the melting of a spherically symmetric nanoparticle, Font and Myers [53] consider the nondimensional system described by Eqs. (4.34, 4.36) subject to (4.38, 4.39, 4.40), where the Stefan condition includes kinetic energy and melting point depression, $T_I < T_m$. The problem configuration is depicted in Fig. 4.1. They employed a standard Gibbs-Thomson relation with curvature $\kappa = 2/R$ to define the interface temperature:

$$T_I(t) = -\frac{\Gamma}{R(t)}, \qquad \text{where} \quad \Gamma = \frac{2\sigma_{sl} T_m}{R_0 \rho_s L_m \Delta T}. \tag{5.90}$$

This is the nondimensional form of (5.4), which demonstrates that the melt temperature decreases with decreasing radius. The parameter Γ indicates the effect of melting point depression; if $\Gamma = 0$, then the interface temperature is fixed at the bulk value. This expression was preferred over the generalised Gibbs-Thomson equation (5.5) for the following reason. The ratio of specific heats is denoted $c = c_s/c_l$. In Table 1.1, we see that $c < 1$ for all materials. A common simplification is to set $c = 1$, which eliminates the term involving $(c_s - c_l)$ in (5.5), and since $p_s - p_l$ is small compared to other terms, the classical Gibbs-Thomson equation (5.4) is obtained. The effect of this approximation on nanoparticle melting was investigated in [51, 82] where it was shown that for large particles, with $R_0 = 100\,\text{nm}$, setting $c = 1$ made a negligible difference to melt times, whereas for smaller particles, $R_0 = 10\,\text{nm}$, a difference of the order 10% was observed. To focus on the effects of density change, we will set $c = 1$ in the following work since it considerably simplifies the analysis and results in relatively small errors.

In [51, 53], the spherical model is applied to the melting of gold nanoparticles. Gold heated to only 10 K above the bulk melt temperature has a Stefan number $\beta \approx 40$. This large value suggests we should immediately rescale time such that $t = \beta\tau$. Then, denoting $\varepsilon = 1/\beta$, the following system is obtained from the equations in Sect. 4.4:

$$\varepsilon\frac{\partial T}{\partial \tau} = \alpha\frac{1}{r^2}\frac{\partial}{\partial r}\left(r^2\frac{\partial T}{\partial r}\right), \qquad \varepsilon\left(\frac{\partial \theta}{\partial \tau} + v_l\frac{\partial \theta}{\partial r}\right) = \frac{1}{r^2}\frac{\partial}{\partial r}\left(r^2\frac{\partial \theta}{\partial r}\right), \tag{5.91a}$$

$$\rho\left[1 + \varepsilon^3\delta\left(\frac{dR}{d\tau}\right)^2\right]\frac{dR}{d\tau} = k\frac{\partial T}{\partial r} - \frac{\partial \theta}{\partial r}, \tag{5.91b}$$

where

$$\rho = \frac{\rho_s}{\rho_l}, \quad \beta = \frac{L_m}{c_l\Delta T}, \quad c = \frac{c_s}{c_l}, \quad \delta = \frac{\rho_l\left(\rho^2 - 1\right)\alpha_l^3}{2R_0^2k_l\Delta T}, \quad k = \frac{k_s}{k_l}. \tag{5.91c}$$

The solid heat equation holds for $r \in [0, R]$ and the fluid equation for $r \in [R, R_b]$. For a fixed boundary temperature, these are subject to

$$\theta(R_b, t) = 1, \quad T(R, t) = \theta(R, t) = -\frac{\Gamma}{R}, \quad \left.\frac{\partial T}{\partial r}\right|_{r=0} = 0, \quad \theta(r, 0) = -\Gamma, \tag{5.91d}$$

and the initial condition for the melt front is $R(0) = 1$. The position of the outer boundary R_b is related to R by Eq. (4.38) and the liquid velocity v_l by (4.39):

$$R_b = \left(\rho - R^3(\rho - 1)\right)^{1/3}, \qquad v_l(r, t) = (1 - \rho)\frac{R^2}{r^2}\frac{dR}{dt}. \tag{5.92}$$

A similar problem employing a Robin condition is studied in [132].

Although we follow the approach of [53], there is a slight difference in the above kinetic energy term and that of [53]. Their Stefan condition is inferred from the Cartesian form provided in the textbook of Alexiades and Solomon [4]. Myers et al. [105] point out an error in the derivation of this version, which was subsequently employed not only in [53] but also in [132, 137]. This was discussed previously in Sect. 4.2.2. The correct form is that given in the above system.

Somewhat contrary to convention, a term of $\mathcal{O}(\varepsilon^3)$ has been retained in the above Stefan condition. In general, this cubic term is expected to be negligible compared to the linear terms. However, as we have frequently seen, and has been shown in a number of studies on nanoparticles, as the particle size tends to zero, the melt rate tends to infinity [82, 157]. The thesis of Back [8] contains an in-depth investigation of finite time blow-up in nanoparticle melting as $R \to 0$. On the other hand, the final time behaviour is neglected in the studies of [51, 53, 132] based on the argument that the equations do not hold beyond the continuum limit of approximately 2 nm, although they do observe large velocities leading up to this limit. With a fixed temperature boundary condition, the initial velocity is infinite. For these reasons, the contribution of kinetic energy is expected to be important at various stages in nanoparticle phase change.

This discussion should give away the fact that we now intend to employ a perturbation series in terms of ε. The retention of an ε^3 term does not invalidate the expansion in any way: in general, it makes a negligible contribution to the results and so its retention is irrelevant. It merely causes extra work in the analysis, but it is retained to capture periods where rapid change occurs. The $\mathcal{O}(\varepsilon^2)$ terms do not affect the small time calculation, so in [53], to prevent the expressions from getting out of hand, only this single cubic term is retained beyond first order.

Writing

$$T = T_0(r,\tau) + \varepsilon T_1(R,\tau) + \mathcal{O}(\varepsilon^2) \tag{5.93a}$$

$$\theta = \theta_0(r,\tau) + \varepsilon \theta_1(R,\tau) + \mathcal{O}(\varepsilon^2), \tag{5.93b}$$

the leading order solution to the above system is

$$T_0 = -\frac{\Gamma}{R}, \qquad \theta_0 = -\frac{\Gamma}{R} + \frac{R_b}{R}\frac{(r-R)}{r}\left(\frac{R+\Gamma}{R_b-R}\right), \tag{5.94}$$

which satisfy $\theta_0(R_b,\tau) = 1$, $T_0(R,\tau) = \theta_0(R,\tau) = -\Gamma/R$, $\partial T_0(0,\tau)/\partial r = 0$. The first-order system is

$$\frac{\partial \theta_0}{\partial \tau} - (\rho - 1)\frac{R^2}{r^2}\frac{dR}{d\tau}\frac{\partial \theta_0}{\partial r} = \frac{1}{r^2}\frac{\partial}{\partial r}\left(r^2\frac{\partial \theta_1}{\partial r}\right), \tag{5.95}$$

$$\frac{\partial T_0}{\partial \tau} = \frac{k}{\rho}\frac{1}{r^2}\frac{\partial}{\partial r}\left(r^2\frac{\partial T_1}{\partial r}\right), \tag{5.96}$$

with boundary conditions $\theta_1(R_b,\tau) = \theta_1(R,\tau) = T_1(R,\tau) = T_{1r}(0,\tau) = 0$. The time derivatives come from the leading order solutions (5.94). The appropriate first-order solutions are

$$\theta_1 = \frac{(\Gamma + R_b)(R_b - r)(r - R)}{6(R_b - R)^2 r}\left\{\frac{3(\rho - 1)R(\Gamma + R)(R_b - R)}{(\Gamma + R_b)r}\right. \tag{5.97}$$

$$+\frac{(\rho-1)R(\Gamma+R)[3-R(R_b+r+R)]}{R_b^2(\Gamma+R_b)}-r+2R_b-R\Bigg\}\frac{dR}{d\tau},$$

$$T_1=-\frac{\rho}{6k}\frac{\Gamma}{R^2}(R^2-r^2)R_\tau. \quad (5.98)$$

The temperatures depend on the single unknown $R(\tau)$. To determine this, we substitute the temperature expressions into the Stefan condition (5.91b) to provide a cubic equation for R_τ subject to the initial condition $R(0) = 1$. Note, since R_b can be expressed in terms of R via Eq. (4.38), this is a single ordinary differential equation for $R(\tau)$. In order to solve the cubic ODE, a variable $z = dR/d\tau$ may first be introduced so that it can be expressed as a cubic polynomial of the form $z^3 + k_1 z + k_2 = 0$. In all the cases tested in [53], this equation had only one real root, $z_1 < 0$, which was then integrated numerically.

So, the perturbation method can reduce the original problem, specified by two partial differential equations, a first-order ordinary differential equation and an algebraic equation, to a single cubic first-order ordinary differential equation. It is possible to proceed further by expanding R and R_b, but this makes the calculation more complex. The current form is readily solved numerically.

The case of equal densities, $\rho_s = \rho_l$, is retrieved by setting $\rho = 1$, $\delta = 0$, and $R_b = 1$. In this case, the Stefan condition reduces to a more manageable form:

$$\left(1+\frac{\varepsilon}{3R}\right)\frac{dR}{d\tau}+\frac{\Gamma+R}{R^2(1-R)}=0, \quad (5.99)$$

with solution:

$$-(1-R^3)+\frac{3(\Gamma+1)-\varepsilon}{2}(1-R^2)$$
$$+(\Gamma+1)(3\Gamma-\varepsilon)\left[(R-1)-\Gamma\ln\left(\frac{\Gamma+R}{\Gamma+1}\right)\right]=3\tau. \quad (5.100)$$

In Sect. 2.2.3, the difficulty of starting a numerical scheme was discussed and the advantage of calculating a small time solution to provide an initial condition. Here we see an issue arising with the temperature expressions for θ_0, θ_1 (Eqs. (5.94, 5.97)); both contain the factor $1/(R_b - R)$. Since $R(0) = R_b(0) = 1$, this indicates an initial singularity (an inevitable result of the fixed temperature boundary condition). So now we seek a small time solution by rescaling time as $t = \omega\bar{t}$, where $\omega \ll 1$. Using a form of co-ordinate transformation similar to those described in Sect. 2.2.1

$$\eta=\frac{r-R}{R_b-R} \quad \text{for} \quad R<r<R_b, \qquad \xi=\frac{r}{R} \quad \text{for} \quad 0<r<R \quad (5.101)$$

and writing $\theta(r,t) = F(\eta,t)\; T(r,t) = G(\xi,t)$ transforms the Stefan condition into

$$\rho(R_b - R)\left[1 + \varepsilon^3\delta\left(\frac{dR}{d(\omega\bar{t})}\right)^2\right]\frac{dR}{d(\omega\bar{t})} = \tag{5.102}$$

$$k\frac{(R_b - R)}{R}\left.\frac{\partial G}{\partial \xi}\right|_{\xi=1} - \left.\frac{\partial F}{\partial \eta}\right|_{\eta=0} .$$

The initial condition, $R(0) = 1$, indicates that for small times, $R \approx 1$ should take a form

$$R = 1 - \lambda(\omega\bar{t})^p, \tag{5.103}$$

where p and λ are constant. The relation between R_b, R, Eq. (4.38), indicates $R_b \approx 1 + (\rho - 1)\lambda(\omega\bar{t})^p$ and so $(R_b - R) \approx \rho\lambda(\omega\bar{t})^p$ and equation (5.102) may now be written as

$$-\rho^2\lambda^2 p(\omega\bar{t})^{2p-1}\left[1 + \varepsilon^3\delta\left(\lambda p(\omega\bar{t})^{p-1}\right)^2\right] = k\rho\lambda(\omega\bar{t})^p\left.\frac{\partial G}{\partial \xi}\right|_{\xi=1} - \left.\frac{\partial F}{\partial \eta}\right|_{\eta=0} .$$

From a physical standpoint, the melting is driven by the temperature gradient in the liquid, F_η. This causes the interface motion and so one, or both of the terms on the left-hand side, must balance the F_η term. Since $\omega \ll 1$, this requires one of the powers $2p - 1$ or $4p - 3$ to be zero (hence removing one of the ω terms). In other words, $p = 1/2$ or $p = 3/4$. In the case of no density jump, $\delta = 0$; then there is only one possibility, namely, $p = 1/2$. However, when $\delta \neq 0$, the second term is largest (consistent with kinetic energy dominating as $t \to 0$), indicating $p = 3/4$. This means that for small times, the radius decreases as

$$R \approx \begin{cases} 1 - \lambda_1 t^{3/4} & \text{if } \delta \neq 0 \\ 1 - \lambda_2 t^{1/2} & \text{if } \delta = 0 . \end{cases} \tag{5.104}$$

The corresponding velocities take the form $R_t \sim t^{-1/4}$, $t^{-1/2}$ for $\delta \neq 0$ and $\delta = 0$, respectively. Both solution forms indicate an infinite velocity as $t \to 0$, but the decrease in radius is faster with no density change (which then results in faster melting). Note, we have not yet determined the constants λ_1, λ_2. They may be determined via the thermal problem.

In both [51, 132], the nanoparticle melting model is solved numerically (as well as finding approximate analytical solutions). Both follow the same route, working on the fixed domain with F, G as defined above, but in addition, they make the change $F \to F/r$, $G \to G/r$ to convert the equations to the Cartesian form. The above small time solutions for R were then introduced to the transformed heat equations to determine analytical expressions for the early time temperatures. In Fig. 5.8, we

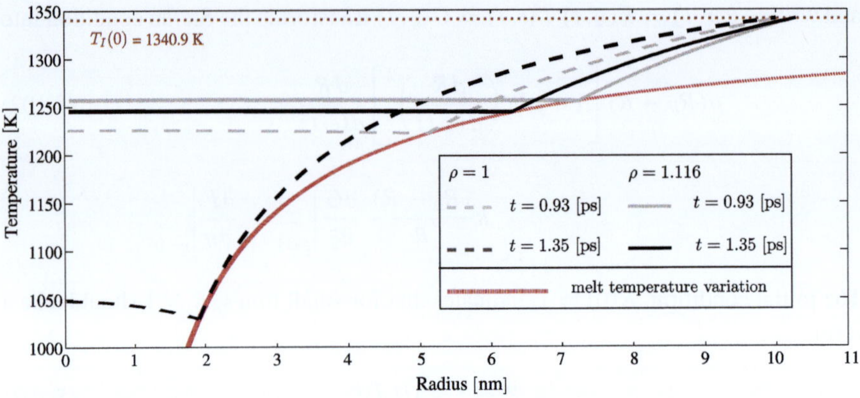

Fig. 5.8 Dimensional temperature profiles for two times during the melting of a gold nanoparticle with $\beta = 100$ and initial particle radius 10 nm. The solid line represents the temperature for $\rho = 1.116$ and the dashed line $\rho = 1$ (light to dark as t increases). The red line shows the melt temperature variation (5.4) and the horizontal dotted brown line shows the initial melt temperature

show a result taken from [53] for the melting of a gold nanoparticle with initial radius $R_0 = 10$ nm and $\beta = 100$ (and hence an initial melt temperature around 1341 K) at times $t = 0.93, 1.35$ ps. The solid lines use the correct density jump for liquid and solid gold, $\rho = 1.116$. The dashed lines correspond to the model with $\rho = 1$. What is clear is that the model with $\rho = 1.116$ exhibits much slower melting – for $t = 1.35$ ps, the solid line indicates $R \approx 6.2$ nm, and the dashed line for $\rho = 1$ has $R \approx 1.9$ nm. The physical mechanism behind the slower melting is easily explained by considering the energy in the system. Melting occurs due to heat being input at the boundary R_b. When $\rho = 1$, this energy goes solely to heating up the material and driving the phase change. However, when $\rho = 1.116$, the fluid must move due to the expansion (or contraction for a material with $\rho_s < \rho_l$) caused by the phase change. This provides another energy sink, namely, kinetic energy, which then results in less energy available to melt the material. Indeed, the small time solution already demonstrated that melting was slower with $\rho > 1$. Mathematically, we may understand the decrease in melt rate through the form of Eq. (4.20) for $c_l = c_s$:

$$\rho_s \left[L_m + \frac{1}{2} \left(\left(\frac{\rho_s}{\rho_l} \right)^2 - 1 \right) \left(\frac{\partial s}{\partial t} \right)^2 \right] \frac{\partial s}{\partial t} = Q \,. \tag{5.105}$$

With $\rho_s > \rho_l$ this may be written

$$\frac{\partial s}{\partial t} = \frac{Q}{\rho_s \left[L_m + f^2(t) \right]} \,, \tag{5.106}$$

where $f^2(t) \geq 0$. This indicates a reduction in s_t.

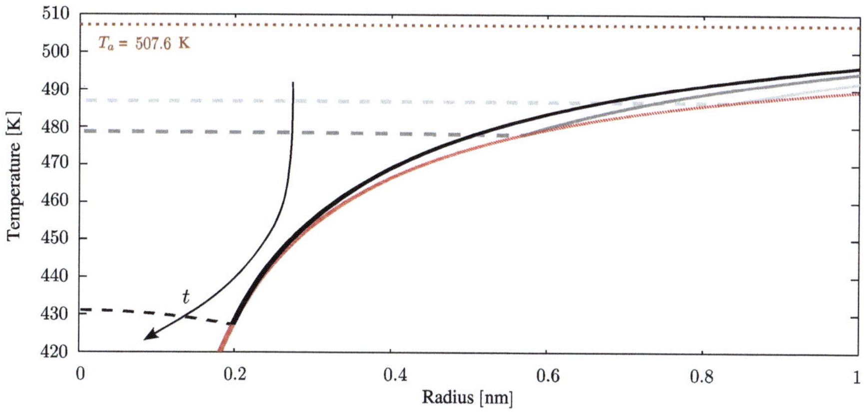

Fig. 5.9 Dimensional temperature profiles for two different times during the melting of a tin nanoparticle for $\beta = 100$ and initial particle radius 10 nm. Relevant parameter values are provided in Table 1.1. The solid and dashed lines represent the temperatures in the liquid and solid, respectively, light to dark as t increases. The red line is the melting temperature given by the generalised Gibbs-Thomson equation (5.5). The horizontal dotted brown line denotes the external temperature, $T_a = 507.6$ K; see Eq. (1.5)

In Fig. 5.9, we present the temperature profiles during the melting of a tin nanoparticle at $t = 3.7, 7.41, 9.08$ ps, again with $R_0 = 10$ nm, $\beta = 100$ (which now corresponds to an initial melt temperature of 490 K). This result is taken from [132], which uses a Robin boundary condition, as well as the added complications of the use of size-dependent latent heat and surface tension and the generalised Gibbs-Thomson relation (5.5).

An interesting feature of both graphs is that the temperature within the solid is greater than the melt temperature, and this becomes more apparent as time increases. This occurs because the melt temperature decreases faster than the solid can diffuse the temperature change. In standard situations, where T_I is constant, the solid acts to slow down melting ($T \leq T_I$, $T_r(s, t) > 0$). At the nanoscale, the opposite can occur: the temperature in the solid phase acts to speed up the melting $T \geq T_I$, $T_r(s, t) < 0$). This may explain the sudden disappearance of particles observed in the experiments of [70]. This unusual temperature gradient behaviour was pointed out in [82] and attributed to superheating, where the solid is everywhere above the (variable) melt temperature. However, other authors define superheating as a surface effect that refers to heating a solid above its bulk melting temperature [139, 161]. If we think of the melt temperature depression as representing the decrease in energy required for the solid particles to move to the liquid phase, then it is clear that the bulk particles are not affected and so should not be described as superheated.

5.4.1 Nanowire Melting

Florio and Myers [49] carry out a perturbation analysis on the melting and solidification of gold nanowires and nanospheres. The melt temperature is described by the standard Gibbs-Thomson relation, Eq. (5.4). A particularly interesting feature of their work concerns the use of Dirichlet and Robin boundary conditions. However, in a great deal of research and indeed in this book, we have worked with the Dirichlet condition in the knowledge that it is physically unrealistic. We quietly justify this by assuming that for sufficiently large heat transfer coefficient $\mathscr{H}$, the results will coincide with those of the Robin condition. In [49], it is pointed out that there are physical limits to the heat transfer: too much energy, for example, with an immediate switch to a new temperature would simply vaporise the material. They define the maximum heat transfer coefficient (before vaporisation occurs) as

$$\mathscr{H}_{max} = \frac{q_{max}}{\Delta T} \quad \text{where } q_{max} = \rho_s u_s v_s \,, \tag{5.107}$$

with u_s the internal energy of the solid and v_s the speed of second sound in the material; see [69]. A maximum heat flux is a requirement for thermodynamic stability. The speed of second sound may be related to phonon velocity $v_s = v_p/\sqrt{3}$ and $v_p = \sqrt{B/\rho_s} \approx 3000$ m/s (where B is the bulk modulus – a measure of the resistance to compression). Under constant pressure, the solid's internal energy may be approximated by the enthalpy h $= c_s \Delta T$; see Eq. (1.12). For gold, this leads to a maximum value $\mathscr{H}_{max} = \rho_s v_s c_s \approx 4.4 \times 10^9$ W/(m^2 K).

In Fig. 5.10, a nondimensional result taken from [49] is presented. It shows the evolution of the melt front for a gold nanowire with $R_0 = 100$ nm, $\beta = 100$. The solid line represents the Robin condition using $\mathscr{H}_{max}$ and the dashed line the Dirichlet condition. Calculations are stopped close to the point that the velocity $R_t < 0$ becomes infinite, where the particle rapidly reduces in size and continuum theory no longer holds. A key feature with the Robin condition is that the initial infinite velocity no longer occurs. Even when the maximum possible value for the

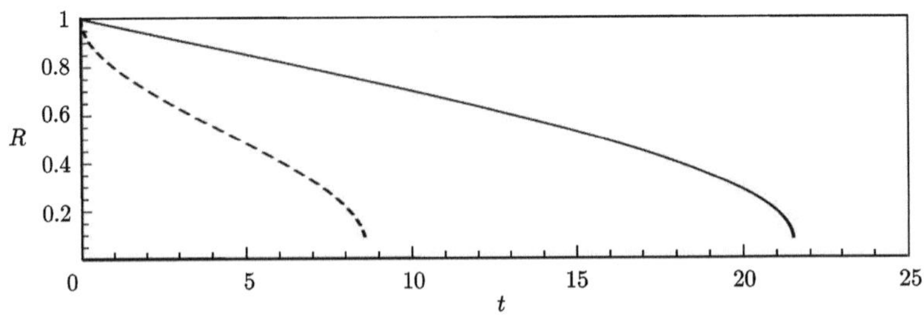

Fig. 5.10 Comparison of the location of the solid-liquid interface as a function of time for $R_0 = 100$ nm, $\beta = 10$ during the melting of a gold nanowire with a Robin boundary condition using $\mathscr{H}_{max}$ (solid), defined by Eq. (5.107), and a fixed temperature boundary condition (dashed); see [49]

heat transfer coefficient is employed, the Dirichlet condition leads to melting at more than twice the rate. The issue may be traced to the value of the Nusselt number arising in the nondimensional analysis $Nu = \mathscr{H}_{max} R_0/k_l \approx 1.72$ for a 100 nm wire. From this, we may deduce that the Dirichlet condition may coincide with the large heat transfer coefficient limit at the macro-scale (when R_0 is sufficiently large) but *at the nanoscale,* $R_0 = \mathscr{O}(10)$ *nm, the Dirichlet condition cannot provide accurate results*. This conclusion was subsequently confirmed in [132].

5.5 Conclusions

In this chapter, we have attempted to showcase the mathematical complexities of a Stefan problem as well as the modelling possibilities, all in situations where the phase change temperature is a variable.

The analysis of one-dimensional supercooling demonstrated how employing a variety of techniques such as perturbations coupled to similarity reductions and small and large time solutions can identify limiting behaviour and so shed light on an important physical process.

The study of nanoparticle melting demonstrated how the temperature in the solid can lag behind the changes in the melt temperature. The fact that the interface temperature can be lower than the temperature in the surrounding phases explains why melting is faster than in a standard problem and, specifically, why nanoparticles disappear at the end of the melting process. Results also indicated that, at least at the nanoscale, kinetic energy can play an important role during melting.

The model for nanocrystal growth involved quite basic mathematics, but it led to the first explicit solution for the growth of the radius in time as well as explaining how to interpret experimental data. Finally, it was shown how the model can help guide the practical process of mass production of nanocrystals. This is an example where the applied mathematician can make a telling contribution – a simple but rigorous model provides practically useful results. Mathematicians may be bad at experiments (well I am anyway), but we are ideally placed to analyse and interpret the mathematical models.

Chapter 5 Exercises

1. A one-dimensional supercooled Stefan problem may be defined by Eqs. (5.8). This can be tackled using the TIM (or any other form of integral method).

   Define

$$\theta = a_0 + a_1(\delta - x) + a_n(\delta - x)^n$$

   (a) Determine a_i by applying the boundary conditions:

$$\theta(s,t) = T_I(t)\,, \qquad \theta(\delta,t) = -1\,, \qquad \theta_x(\delta,t) = 0.$$

(b) Apply the heat balance integral to find a differential equation involving s, δ.
(c) The system may be closed through applying the Stefan condition, using the above temperature profile and then defining the linearised interface temperature $T_I(t)$ through $s_t = -T_I(t)$.
From the Stefan condition, we may express δ in terms of s, s_t and so eliminate it (leading to a second-order equation). We are now faced with two coupled ordinary differential equations. Solve these numerically for appropriate nondimensional numbers, using the initial conditions $s(0) = 0$, $s_t(0) = 1$ and with the optimal value $n = 3.57$. Also try $n = 2$; how do the solutions compare?
Note, the condition $s_t(0) = 1$ leads to an initial singularity in the acceleration; a small time solution of the form $s_t = 1 + At^\alpha$ may be required.
The nonlinear case is dealt with in [52]. Feel free to try this.

2. Compare the large time velocities, s_t, for the solidification of a supercooled melt subject to linear and nonlinear kinetic undercooling for copper and salol, with values of $\beta < 1$. Investigate what happens as $\beta \to 1^-$.
3. The variation of interface temperature with curvature is expressed by the generalised Gibbs-Thomson equation (5.5). Assuming negligible pressure effects, $p_l = p_s$, plot $T_I(R)$ for gold and water spheres. Compare this to the result of the standard Gibbs-Thomson equation (5.4). Note, it is simpler to plot $R(T_I)$, for say $T_I \in [0.1T_m, 0.95T_m]$ (it differs in each case). From this, you will see the order of the radius where this effect is important.
In each case, at what radius should the generalised Gibbs-Thomson relation be assumed to fail?
4. Consider the dimensional single crystal growth system:

$$\frac{\partial c}{\partial t} = \frac{D}{r^2}\frac{\partial}{\partial r}\left(r^2 \frac{\partial c}{\partial r}\right), \qquad r > r_p(t)$$

$$\frac{dr_p}{dt} = V_m D \left.\frac{\partial c}{\partial r}\right|_{r_p},$$

subject to

$$c|_{r_p+\delta} = c_b(t), \quad c(r_p, t) = S + \left.\frac{D}{k}\frac{\partial c}{\partial r}\right|_{r=r_p}, \quad c(r, 0) = c_b(0), \quad r_p(0) = r_{p0}.$$

(a) Determine the pseudo-steady concentration $c(r, t)$ and consequently the growth rate dr_p/dt.
(b) Simplify the expressions by taking the limit $\delta \gg r_p$.

(c) Now let $S = S_{eq} = S_{\infty}e^{\alpha_c/r_m}$, where r_m is the maximum radius achieved during the process and substitute for

$$c_b(t) \approx c_0 - \frac{4\pi N_0}{3V_M}{r_p}^3,$$

to obtain an equation of the form

$$\frac{dr_p}{dt} = \frac{Dk}{D + kr_p}(a^3 - b^3 r_p^3).$$

Verify that this matches the nondimensional form of (5.83) (when $S = S_{eq}$).

5. Equation (5.84) defines the growth of a nanocrystal radius with time

$$r_p = \frac{r_m}{2}\frac{\left[1 + 2f(r_{ps})\exp\left(\frac{t-t_s}{G}\right) - \sqrt{-3 + 12f(r_{ps})\exp\left(\frac{t-t_s}{G}\right)}\right]}{\left[-1 + f(r_{ps})\exp\left(\frac{t-t_s}{G}\right)\right]}.$$

In Sect. 2.1.4, we obtained a solution for the one-phase growth into an infinite solution. The similarity solution suggested $r_p = 2\lambda\sqrt{t}$. Is there any resemblance between the two solutions? Could the similarity solution be employed as a simple approximate solution? (Work with the data of [149], which gives $G = 789$, $r_m = 1.24$ nm, $r_{ps} = 1.16$ nm, $t_s = 720s$).

6. Replace the Dirichlet boundary condition $\theta(R_b, t) = 1$ in problem (5.91) by the Newton cooling condition (4.39) and find the leading order solution (assuming the expansion in ε given by (5.93)):

(a) Compare the solution of the model with the temperature profiles (5.94) and the position of the front (5.100) (setting $\varepsilon = 0$ in the latter).

(b) Make a plot of $R(t)$ versus t for increasing values of Nu (e.g., 1, 5, 10, 50) to prove that the leading order problem using Newton cooling converges to (5.100) as Nu increases.

7. The large Stefan number symmetric melting of a wire may be described to leading order in $\varepsilon = 1/\beta$ by

$$0 = \frac{1}{r}\frac{\partial}{\partial r}\left(r\frac{\partial T}{\partial r}\right) = \frac{1}{r}\frac{\partial}{\partial r}\left(r\frac{\partial \theta}{\partial r}\right),$$

$$\rho\frac{dR}{d\tau} = k\frac{\partial T}{\partial r} - \frac{\partial \theta}{\partial r},$$

where $\rho = 1$. For a fixed boundary temperature, these are subject to

$$\theta(R_b, t) = 1, \quad T(R, t) = \theta(R, t) = -\frac{\Gamma}{R}, \quad \left.\frac{\partial T}{\partial r}\right|_{r=0} = 0, \quad \theta(r, 0) = -\Gamma,$$

where $R_b = 1$ (for the case $\rho = 1$) and the initial condition for the melt front is $R(0) = 1$:

(a) Solve the above system to determine $R(t)$ (or $t(R)$).
(b) Now determine the Stefan condition for the case $\rho \neq 1$. This will require determining the position $R_b(t)$, which requires imposing an incompressibility condition within the liquid phase as well as Rankine-Hugoniot to the mass balance.
The ODE can only be solved numerically. Try it and compare solutions for $\rho < 1, > 1, = 1$.

Chapter 6
Non-Fourier Stefan Problems

The breakdown of Fourier's law, at both small length- and time-scales, has been predicted theoretically, demonstrated via molecular dynamics and observed experimentally; see [58, 140]. Non-Fourier effects have even been observed in samples with length-scales as large as 10 microns – far beyond the nano-regime [74]! At the nanoscale, this is partially attributed to the increased ratio of surface-to-bulk atoms, which introduces a size dependence to key thermodynamic parameters such as latent heat and surface energy as well as the phase change temperature investigated in Sects. 5.1, 5.4. A second reason is that the method of energy transport appears different when moving from the macroscale to the nanoscale. In solids, energy transport may be described by the action of phonons.[1] At the nanoscale, thermal energy transport may be viewed as a ballistic process driven by infrequent, random collisions between phonons. However, at the macroscale, due to the large number of phonons and consequent frequent collisions, the process may be viewed, on average, as diffusive. Diffusive transport of heat is well described by Fourier's law, which is based on a continuum assumption (so there are no individual phonon collisions, just the collective result); an unfortunate consequence of this is the prediction of an infinite speed of heat propagation. If the finite time between phonon collisions is taken into account, then, as we will see in this chapter, a wavelike propagation of heat, with finite speed, is found, which is not described by Fourier's law. Jou's book [69] on nonclassical heat transport (nonequilibrium thermodynamics) provides an interesting introduction to the subject.

There are many different models that aim to correctly describe heat transfer in situations where Fourier's law is invalid; these may be categorized into micro-, meso-, and macroscopic approaches. Micro- and mesoscopic models focus on each of the heat carriers or on their distribution, whereas the macroscopic models aim to describe the system in terms of quantities such as the temperature or the heat flux.

[1] Simon [141] states that a phonon may be thought of as a particle or as a quantised wave – a concept more familiar in the wave-particle duality of light and specifically photons.

T. G. Myers, *Theoretical and Practical Stefan Problems*, Interdisciplinary Applied Mathematics 63, https://doi.org/10.1007/978-3-032-04826-4_6

The computational cost of approaches belonging to the first two categories, such as molecular dynamics, the Boltzmann transport equation (BTE), or the equation of phonon radiative transfer, is high, and therefore only very small systems are usually considered.

The unphysical infinite speed of heat propagation prompted Cattaneo [30] to propose that a temperature gradient can only induce a thermal flux after a finite amount of time has passed and so introduced a time lag. The resultant model, termed the Maxwell-Cattaneo (MC) equation or relativistic heat equation, extended Fourier's law by incorporating a term to account for the time lag, leading to a flux dependent on the history of the temperature gradient, which is thus referred to as a memory term. (Maxwell got his name in there since the MC equation is identical to the Telegrapher's equation, which is derived from Maxwell's equations of electrodynamics. So why not the Cattaneo-Telegrapher equation? Maxwell is famous enough already!) Starting from the linearised Boltzmann transport equation, Guyer and Krumhansl developed a model including both nonlocal and memory effects. In the limit as the nonlocal effects tend to zero (e.g. when the phonon mean free path tends to zero or is small in comparison to some characteristic length-scale), the MC equation is retrieved. Although the Guyer-Krumhansl (GK) model was originally derived for situations with extremely low temperatures, recent studies have shown that it may be accurate at much higher temperatures: Guo and Wang [58] demonstrate good agreement with experimental data at $T = 300$ K. A particular feature of the GK model is the similarity to the Navier-Stokes equations, which then enables non-Fourier heat flow to be conceptualised in terms of fluid mechanics and thus permits knowledge transfer between the two fields.

In this chapter, we begin with a simple analysis of the GK equations to describe heat flow in a cylindrical nanowire. A key feature here is to provide a suitable model for the thermal conductivity. At the macroscale, this is generally accepted as an intensive property (i.e. one that does not depend on the system size or the amount of material in the system), but it turns out to be an extensive property at the nanoscale. Once the GK equation has been established as a reliable descriptor of heat flow at the nanoscale, we extend the equations to deal with a one-dimensional phase change problem. The analysis will closely follow that of Calvo et al. [27, 62, 63, 119]. In line with these works, the discussion will be in terms of melting at the nanoscale, but the models work equally well in the context of ultrafast heating.

6.1 Mathematical Model for Heat Flow in a Nanowire

Here we follow the work of Calvo et al. [26] to determine an expression for the *effective thermal conductivity* (ETC) in a nanowire. A basic form of GK model may be written as

$$\rho_s \frac{\partial u_s}{\partial t} = -\nabla \cdot \mathbf{q}, \tag{6.1}$$

$$\tau_R \frac{\partial \mathbf{q}}{\partial t} + \mathbf{q} = -k\nabla T + \ell^2(\nabla^2 \mathbf{q} + 2\nabla\nabla \cdot \mathbf{q}) , \tag{6.2}$$

where T is the temperature and u_s the internal energy (related to temperature via Eq. (4.14a)). The heat flux vector is denoted $\mathbf{q}$, and assuming axial symmetry, we may write $\mathbf{q} = (v, w)$ to distinguish the components in the r, x directions. The two unfamiliar parameters are as follows:

1. τ_R, which represents the thermal resistive relaxation time, that is, the mean time between collisions among the heat carriers (phonons) that do not conserve their momentum (and so transfer energy).
2. ℓ is the phonon mean free path, which is the mean distance that a phonon travels between collisions.

The first term in (6.2) is known as the Cattaneo memory term, and it is dominant on time-scales comparable to τ_R. The ℓ terms represents nonlocal effects that dominate heat transport in systems with length-scales comparable to ℓ. In the limit $\ell \to 0$, the Maxwell-Cattaneo equation is retrieved. In the limit $\tau_R, \ell \to 0$, Fourier's law is retrieved. The thermal conductivity $k = k(T)$ in Eq. (6.2) is the bulk value at some temperature T (by bulk here we mean in a large sample). However, in experiments at the nanoscale, the measured effective thermal conductivity (ETC) may differ significantly from this bulk value. The ETC is defined by

$$-k_{\text{eff}} \frac{\partial T}{\partial x} = Q \tag{6.3}$$

and Q is the heat flux per unit area.

Comparison with the incompressible Navier-Stokes equations presented in Sect. 3.1 demonstrates an obvious analogy (if a source term is added) if $\mathbf{q}$ is equated with the fluid velocity $\mathbf{u}$, while temperature is equated to pressure. From Eq. (6.2), we may also identify a "thermal viscosity" $\mu = \ell^2/k$. Fluid flow may be driven by a pressure gradient; here temperature gradient drives the heat flow. In the context of heat flow in a solid, the pressure in the internal energy expression (4.14a) is constant, and so the derivative of internal energy, required in (6.1), is $\partial u_s/\partial t = c_s T$. Equations (6.1, 6.2) then form a coupled pair of partial differential equations for the temperature and heat flux.

Since the goal of this section is to determine the ETC in a nanowire, we now consider a situation where the heat flow is driven by a difference in temperatures applied at the ends of a circular cross-section wire of constant radius R:

$$T|_{x=0} = T_0 , \qquad T|_{x=L} = T_1 , \tag{6.4}$$

where $T_0 > T_1$. Analogous to a pressure-driven fluid flow, suitable heat flux boundary conditions are

$$v|_{r=0} = \left.\frac{\partial w}{\partial r}\right|_{r=0} = 0 , \qquad v|_{r=R} = 0 . \tag{6.5}$$

The first two conditions result from symmetry along the centre-line of the wire; the final condition indicates that there is no heat flow perpendicular to the outer boundary (which in the experimental results described later correspond to an insulated outer boundary). A further condition, analogous to the slip condition of fluid mechanics (3.12), is applied at the boundary:

$$w|_{r=R} = -L_s \left.\frac{\partial w}{\partial r}\right|_{r=R}, \tag{6.6}$$

where the "slip length" is proportional to the nonlocal length, $L_s = C\ell$, and C is a dimensionless parameter that quantifies phonon-boundary scattering and surface roughness [138]. In [26], it is chosen as $C = \exp(-R/\ell)$, which matches the expected behaviour in the limits of large and small R/ℓ. The heat flux per unit cross-sectional area, required to calculate the ETC, is

$$Q = \frac{2}{R^2}\int_0^R w(r,x)\, r\, dr\,. \tag{6.7}$$

The following nondimensional variables are now introduced:

$$\hat{r} = \frac{r}{R}, \quad \hat{x} = \frac{x}{L}, \quad \hat{v} = \frac{v}{v_0}, \quad \hat{w} = \frac{w}{w_0}, \quad \hat{T} = \frac{T - T_1}{T_0 - T_1}, \quad \hat{k} = \frac{k}{k_0}, \tag{6.8}$$

where $R/L = \varepsilon \ll 1$, v_0, w_0 are (unknown) typical values of the axial and radial components of the heat flux and k_0 is a reference value for the bulk thermal conductivity (which varies with temperature).

In keeping with the fluid analogy, we now recall the reduction of the Navier-Stokes equations consistent with lubrication theory; see Sect. 3.1. A key assumption of lubrication theory is that, due to the thinness of the fluid layer, the main flow must be in the direction of the larger dimension (since there is not much space for movement in the perpendicular direction). In the case of a nanowire, the heat flow must be predominantly along the axial direction; here this corresponds to the x-direction. Consequently, we assume $w_0 \gg v_0$ and will scale accordingly. The steady versions of Eqs. (6.1,6.2) are now

$$\frac{v_0}{R}\frac{1}{r}\frac{\partial}{\partial r}(rv) + \frac{w_0}{L}\frac{\partial w}{\partial x} = 0\,, \tag{6.9}$$

$$v_0 v = -\frac{k_0 \Delta T}{R} k \frac{\partial T}{\partial r} + \frac{\ell^2 v_0}{R^2}\left(\frac{1}{r}\frac{\partial}{\partial r}\left(r\frac{\partial v}{\partial r}\right) + \varepsilon^2 \frac{\partial^2 v}{\partial x^2}\right), \tag{6.10}$$

$$w_0 w = -\frac{k_0 \Delta T}{L} k \frac{\partial T}{\partial x} + \frac{\ell^2 w_0}{R^2}\left(\frac{1}{r}\frac{\partial}{\partial r}\left(r\frac{\partial w}{\partial r}\right) + \varepsilon^2 \frac{\partial^2 w}{\partial x^2}\right), \tag{6.11}$$

where, as usual, the hats have been dropped immediately. Balancing equation (6.9) indicates $v_0 = \varepsilon w_0$. From the final equation, we choose $w_0 = k_0 \Delta T/L$, which then allows the system to reduce to Fourier in the limit $\ell \to 0$. The resultant leading order system (i.e. neglecting terms of $\mathcal{O}(\varepsilon^2)$) is

$$\frac{1}{r}\frac{\partial}{\partial r}(rv) + \frac{\partial w}{\partial x} = 0\,, \qquad \frac{\partial T}{\partial r} = 0\,, \tag{6.12}$$

$$w = -k\frac{\partial T}{\partial x} + \frac{(Kn)^2}{r}\frac{\partial}{\partial r}\left(r\frac{\partial w}{\partial r}\right)\,, \tag{6.13}$$

where the Knudsen number $Kn = \ell/R$ is the ratio of the mean free path length to the physical length-scale (in this case, the radius of the wire). There is clearly a close analogy to the treatment of the above equations to that outlined in the lubrication section Sect. 3.1. The above equations are subject to

$$v|_{r=0} = \left.\frac{\partial w}{\partial r}\right|_{r=0} = 0\,, \qquad v|_{r=1} = 0\,, \tag{6.14}$$

$$w|_{r=1} = -Kn\,C\left.\frac{\partial w}{\partial r}\right|_{r=1}\,, \qquad T(0,r) = 1\,, \qquad T(1,r) = 0. \tag{6.15}$$

The (nondimensional) ETC is defined through the relation

$$-k_{\text{eff}}\frac{\partial T}{\partial x} = Q = 2\int_0^1 w(r,x) r\,dr\,. \tag{6.16}$$

The second of Eqs. (6.12) demonstrates that T is independent of r; (6.13) is then a standard second-order differential equation for $w(r)$, which is easily integrated to give

$$w = -k\left(1 - \frac{I_0(r/Kn)}{I_0(1/Kn) + CI_1(1/Kn)}\right)\frac{\partial T}{\partial x}\,, \tag{6.17}$$

where I_ν is the modified Bessel function of the first kind of order ν. The heat flux per unit area may now be calculated:

$$Q = 2\int_0^1 w(r,x) r\,dr = -k\left(1 - \frac{2Kn I_1(1/Kn)}{I_0(1/Kn) + CI_1(1/Kn)}\right)\frac{\partial T}{\partial x}. \tag{6.18}$$

Hence, from the definition of (6.16), we have obtain a simple expression for the effective thermal conductivity:

$$k_{\text{eff}} = k\left(1 - \frac{2Kn\,I_1(1/Kn)}{I_0(1/Kn) + CI_1(1/Kn)}\right). \tag{6.19}$$

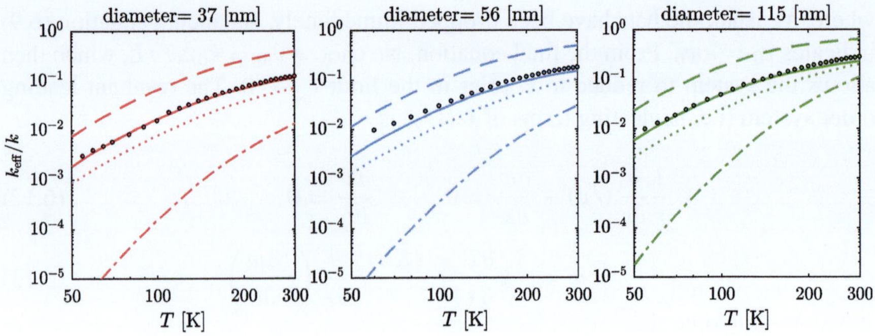

Fig. 6.1 Comparison of theoretical predictions and experimental data for k_{eff} against temperature for silicon nanowires of diameter 37 nm (left figure), 56 nm (middle figure), and 115 nm (right figure). Solid lines correspond to Eq. (6.19); dashed, dotted, and dash-dot lines represent the model predictions given by [5, 43, 77], respectively; circles denote experimental data taken from [74]

Although Q must be constant (provided no heat is lost at the boundary $r = 1$), it may not be possible to integrate equation (6.18) analytically to determine the temperature profile – the bulk conductivity k and C vary with temperature. However, it is simply a first-order ODE of the form $T_x = f(T, Q)$ and hence easily evaluated numerically.

In Fig. 6.1, the ETC predicted by Eq. (6.19) (solid line) is compared with experimental data (circles) for silicon nanowires of diameter 37, 56, and 115 nm, for temperatures between 50 and 300 K, taken from [74]. Also shown are predictions from the theoretical models of [5, 43, 77]. The above analytical expression, Eq. (6.19), clearly provides excellent agreement with the data. The models of [43, 77] also stem from the GK equation, albeit with different boundary conditions. The model of [43] applies a no-slip condition, such that $w|_{r=1} = 0$, whereas in [77], a prescribed flux condition is applied. From the graph, it is quite clear that the former provides a reasonable approximation, showing the correct trend; the latter leads to poor agreement for all T. The result of [5] comes from a linearised Boltzmann equation that does not allow for conditions to be applied at the wire perimeter. It also shows the correct trend, but with a significant offset. Hence, it seems that models based on the GK equation can accurately describe heat flow at the nanoscale, but the choice of boundary condition is crucial to the model's success.

The results presented in Fig. 6.1 clearly show that the GK equation may be used to describe heat flow at the nanoscale (provided correct boundary conditions are applied) as well as solving a fundamental problem. Consequently, in the following section, we will use a GK model to investigate nanoscale phase change, where detailed experimental data is not available.

6.2 Mathematical Model for Phase Change

Aware that the thermal conductivity is size dependent at the nanoscale, Font [50] and Calvo [25] studied one-dimensional phase change governed by Fourier's law but with a size-dependent k, the first paper dealing with a Dirichlet boundary condition and the second with a Robin condition. The two-phase Stefan problem governed by the Maxwell-Cattaneo form has been employed by [63, 142] in Cartesian and spherically symmetric configurations. Hennessy et al. [62] used the GK equation in an asymptotic study of one-dimensional phase change. Below we will follow the model of Calvo et al. [27], which is based on the GK formalism. This deals with one-dimensional solidification, starting from a seed crystal, in a liquid originally maintained at the phase change temperature (so a classic one-phase model with a twist).

The mathematical model follows from the one-dimensional Cartesian version of Eqs. (6.1, 6.2). Substituting for the internal energy, given by Eq. (4.14a) (with constant pressure and melt temperature),

$$\rho_s c_s \frac{\partial T}{\partial t} = -\frac{\partial q}{\partial x} \tag{6.20}$$

$$\tau_R \frac{\partial q}{\partial t} + q = -k \frac{\partial T}{\partial x} + 3\ell^2 \frac{\partial^2 q}{\partial x^2}, \tag{6.21}$$

where q is the heat flux in the x-direction and the factor 3 results from the combination of nonlocal memory terms. The Stefan condition is

$$\rho_s L_m \frac{\partial s}{\partial t} = -q. \tag{6.22}$$

Solidification is driven by imposing a fixed boundary temperature $T(0, t) = T_0$. At the interface, $T(s, t) = T_m$ is constant. Solidification starts from a seed crystal of size s_c such that the initial conditions are

$$s(0) = s_c \qquad T(x, 0) = T_m \qquad q(x, 0) = 0. \tag{6.23}$$

The concept of the seed crystal was introduced in the previous chapter; here it is motivated by two considerations: it avoids the unphysical infinite rate of phase change due to the fixed temperature boundary condition; it is likely to be used in molecular dynamics simulations, and so results obtained by this approach may be compared with the present model.

The system is nondimensionalised in the following manner:

$$\hat{x} = \frac{x}{s_c}, \quad \hat{s} = \frac{s}{s_c}, \quad \hat{t} = \frac{t}{\tau}, \quad \hat{q} = \frac{q}{q_0}, \quad \hat{T} = \frac{T - T_m}{T_f - T_0}, \tag{6.24}$$

where $q_0 = k\Delta T/s_c$, $\tau = s_c^2/\alpha_s$. This results in

$$\frac{\partial T}{\partial t} = -\frac{\partial q}{\partial x} \tag{6.25}$$

$$\gamma\frac{\partial q}{\partial t} + q = -\frac{\partial T}{\partial x} + \eta^2\frac{\partial^2 q}{\partial x^2}\,, \tag{6.26}$$

where $\gamma = \tau_R/\tau$, $\eta = \sqrt{3}\ell/s_c$. The parameter γ is the Cattaneo number, and η is a Knudsen number (but with different notation to distinguish it from the form of the previous section). The Stefan condition is

$$\beta\frac{\partial s}{\partial t} = -q\,. \tag{6.27}$$

These are subject to

$$T(0,t) = -1\,, \qquad T(s,t) = 0\,, \qquad T(x,0) = q(x,0) = 0\,, \qquad s(0) = 1\,. \tag{6.28}$$

A list of physical parameter values for the case of silicon are provided in Table 6.1. From these, we may obtain the nondimensional numbers appearing in the above equations. In Table 6.2, we present values for these numbers appropriate to the solidification of silicon with $\beta = 10$ and for seed crystals of 2, 5, and 10 nm.

It turns out that the above system exhibits a variety of different behaviours depending on the relative size of nonlocal effects, relaxation time, and Stefan number. In the supplementary material of [27], an asymptotic analysis of a number of these limits, over a variety of time-scales, is carried out. They also solve the system numerically, after applying the boundary fixing transformation $\eta = x/s(t)$.

Stating that the stability of the numerical solutions is greatly improved through the use of a small time solution as an initial condition, they examine the limit $t \ll 1$ for the case $\gamma \gg \eta = \mathcal{O}(1)$, that is, heat transport in the seed crystal is initially dominated by memory effects. In this case, the equations must be rescaled such that

Table 6.1 Thermophysical properties of silicon taken from [27]

Thermal conductivity	k	22.1	W/m K
Specific heat	c	1032	J/kg K
Density	ρ	2296	kg/m^3
Thermal diffusivity	α	9.33×10^{-6}	m^2/s
Phase change temperature	T_m	1687	K
Latent heat	L_m	1.787×10^6	J/kg
Relaxation time	τ_R	32.16	ps
Mean free path	ℓ	8.79	nm

Table 6.2 Values of τ, Q_0, γ, η when $\beta = 10$ for three different seed crystal sizes. Taken from [27]

s_c (nm)	τ (ps)	Q_0 (W/mm^2)	γ	η
2	0.43	1.9	75	7.7
5	2.7	0.77	12	3.0
10	11	0.38	3.0	1.5

$t = \varepsilon\bar{t}$, where ε is arbitrarily small; other variables must be scaled as

$$T = \bar{T}\,, \quad s = \bar{s}\,, \quad x = (\varepsilon/\gamma)^{1/2}\bar{x}\,, \quad q = (\varepsilon\gamma)^{-1/2}\bar{q}\,. \tag{6.29}$$

The T, s scalings arise from noting that the initial conditions indicate $T = s = \mathcal{O}(1)$ (so we make no change); the other scalings are chosen to obtain a meaningful balance in the governing equations. With these new scalings, the Stefan condition indicates $\bar{s}_{\bar{t}} = \mathcal{O}(\varepsilon^{1/2})$, and since ε may be chosen arbitrarily small, this indicates $\bar{s} = 1$ as a leading order approximation. That is, we work on a time-scale too fast for the boundary motion to be noticed. The flux and temperature then satisfy the following equations over a fixed domain:

$$\frac{\partial\bar{T}}{\partial\bar{t}} = -\frac{\partial\bar{q}}{\partial\bar{x}}\,, \qquad \frac{\partial\bar{q}}{\partial\bar{t}} = \eta^2\frac{\partial^2\bar{q}}{\partial\bar{x}^2}\,, \tag{6.30}$$

such that $\bar{T}(0,\bar{t}) = -1$, $\bar{T} \to 0$ as $\bar{x} \to \infty$, $\bar{T}(\bar{x},0) = \bar{q}(\bar{x},0) = 0$. This system is easily resolved using a similarity transformation:

$$T = -1 + \mathrm{erf}\left(\frac{x}{2\sqrt{\zeta t}}\right), \qquad q = -\sqrt{\frac{\zeta}{\pi t}}e^{-x^2/(4\zeta t)}\,, \tag{6.31}$$

where $\zeta = \eta^2/\gamma$. This provides the initial temperature and flux at a specified time $t_0 \ll 1$ in the numerical scheme of [27].

In Figs. 6.2 and 6.3, we show results taken from [27] for the evolution of the phase change front and the mean flux $\langle q\rangle$ where

$$\langle q\rangle = \frac{1}{s}\int_0^s q\,dx\,. \tag{6.32}$$

In both cases, the Stefan number $\beta = 10$, which corresponds to a temperature scale $\Delta T = 173.2$K. All other parameter values are specified in Tables 6.1 and 6.2. Figure 6.2 shows the case where the seed crystal has size 2 nm. The left-hand figure shows a comparison of results obtained from a standard Fourier model, the GK model given by Eqs. (6.25, 6.26), and the MC model, obtained by setting $\eta = 0$ in the GK model. The left-hand figure shows the evolution of the phase change front. As can be seen, for silicon, the GK and Fourier models are virtually indistinguishable. On the other hand, the MC model does exhibit a different behaviour. In fact, the MC model has s stationary for the first 10 ps (consistent

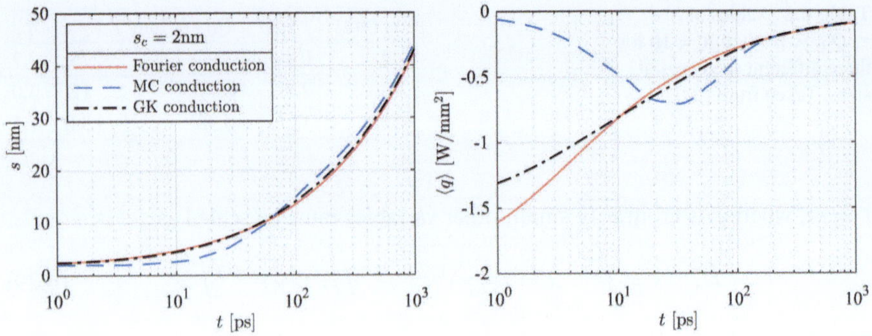

Fig. 6.2 Evolution of $s(t)$ for silicon and the mean flux according to Fourier's law (red, solid lines), the MC equation (blue, dashed lines), and the GK equation (green, dashed-dotted lines) for $\beta = 10$ and $s_c = 2$ nm; see [27]

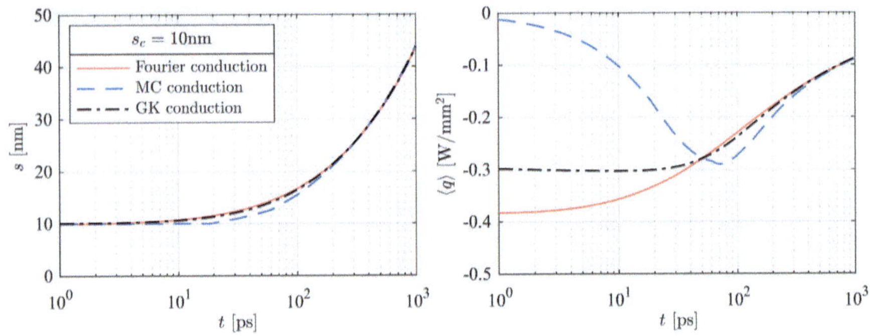

Fig. 6.3 Evolution of $s(t)$ for silicon and the mean flux according to Fourier's law (red, solid lines), the MC equation (blue, dashed lines), and the GK equation (green, dashed-dotted lines) for $\beta = 10$ and $s_c = 10$ nm; see [27]

with the small time solution); it then catches up quickly to the other models, overshoots, and then tends towards Fourier and GK at larger times. Physically, this may be understood from the fact that in the MC model, heat propagates purely as a thermal wave with finite speed, in contrast to the dominant early time behaviour of the Fourier and GK models. Although the interface motion predicted by Fourier and GK is almost identical, the heat flux is not, suggesting that the choice of the underlying model has an important impact on how heat propagates at small times ($<$ 100 ns). The results of Fig. 6.2 suggest that nonclassical mechanisms may be more easily detected by measuring the total heat flux through the solid rather than the progression of the solidification front.

Figure 6.3 shows the equivalent results, but with a 10 nm seed crystal. Again differences between GK and Fourier are really only detectable in the mean flux measurement.

Given the difference in the models, it is quite surprising that GK and Fourier conduction provide such similar results – was the effort of formulating the GK

Stefan problem worthwhile? This was investigated further in [27]. Returning to the dimensional GK system (6.20, 6.21), the final term of (6.21) may be replaced via Eq. (6.20) to give

$$\tau_R \frac{\partial q}{\partial t} + q = -k \left(\frac{\partial T}{\partial x} + 3 \frac{\ell^2}{\alpha} \frac{\partial^2 T}{\partial x \partial t} \right) . \tag{6.33}$$

Recall that τ_R, the resistive thermal relaxation time, for silicon an appropriate value is provided in Table 6.1. In the special case $\tau_R = 3\ell^2/\alpha$, the above equation may be rearranged to

$$3 \frac{\ell^2}{\alpha} \frac{\partial}{\partial t} \left(q + k \frac{\partial T}{\partial x} \right) + \left(q + k \frac{\partial T}{\partial x} \right) = 0 , \tag{6.34}$$

which is obviously satisfied by Fourier's law, $q = -k\partial T/\partial x$. This model reduction, valid for the special case where $\tau_R \approx 3\ell^2/\alpha$, is termed Fourier resonance. From Table 6.1, we obtain a value $3\ell^2/\alpha = 2.48 \times 10^{-11} = 0.77\tau_R$, so the system is indeed close to Fourier resonance. From this, it was concluded that non-Fourier effects could only be detected through observation of the position of the interface when the material and conditions are such that Fourier resonance is not approached. In conditions close to Fourier resonance, non-Fourier effects may only be detected by measurements of the heat flux.

This similarity of results between Fourier and non-Fourier models does not always occur. Hennessy et al. [63] consider the melting of a spherical nanoparticle driven by a heat source at the outer boundary. Kinetic energy was neglected in the Stefan condition and the Gibbs-Thomson relation applied to determine melting point depression. The heat flow was governed by the Maxwell-Cattaneo equation, and in order to focus on the effect of finite-time heat transfer, an average density and heat capacity were imposed over the solid and liquid regions, while thermal conductivity changed between phases. Their results demonstrated that thermal relaxation could either decelerate or accelerate the melting process relative to the case of Fourier conduction. For one case studied in their paper, the melting time decreased by more than a factor of 10. However, the most interesting finding was that when using the MC equation the temperature must have a jump at the interface. The jump condition prevents supersonic melting, where the speed of the melt front exceeds the speed of heat propagation. This is a qualitative difference in behaviour from the Fourier prediction.

6.3 Conclusion

In this chapter, we focussed on the Guyer-Krumhansl equation to describe the flow of heat at the nanoscale. The analogy with fluid mechanics provided a means of knowledge transfer and certainly aided in choosing an appropriate boundary condition in the analysis of nanowire heat flow. The results showed excellent agreement with data for the effective thermal conductivity.

In the second section, the surprising result arose that both Fourier and the GK equations lead to virtually identical motion of the phase change front, despite markedly different forms of heat flow. It appears that this was an unlucky (or lucky) circumstance in that the material and conditions chosen were close to Fourier resonance, where the GK model is satisfied by Fourier's law. However, greater differences may be observed when using different materials or a different heat flow model, such as Maxwell-Cattaneo.

In closing this chapter and the book, it is certainly worth pointing out that the field of non-Fourier heat flow and phase change is currently wide open. Physicists are still striving to understand the basics of heat flow, let alone an accompanying phase change theory. Yet, it is a vital area of research given the ever-increasing miniaturisation of electrical components.

One final comment that applies not only for the topics of this chapter but for almost all the moving boundary problems examined in this book. As the physicists or chemists or engineers develop new experiments and discover new problems, the mathematicians can be waiting to unlock their secrets. Careful mathematical analysis can provide invaluable insight and understanding of physical phenomena, which cannot come from other disciplines. A synergy between scientists of different disciplines is essential to move the boundaries of research.

> *Neglect of mathematics works injury to all knowledge, since he who is ignorant of it cannot know the other sciences or the things of the world.*
>
> Roger Bacon 1214–1292

Chapter 6 Exercises

1. An experimental method for examining non-Fourier modes of heat conduction involves subjecting a sample of material to a high-frequency laser and measuring the surface temperature through changes in the surface reflectivity (this method is called thermoreflectance). This exercise will involve developing and analysing a one-dimensional model of a thermoreflectance experiment. Assume that the position of the irradiated surface is given by $x = 0$, where x denotes the distance from a point in the sample to the surface. The thermal flux applied to the sample by the laser can be modelled through the boundary condition $q(0,t) = q_0 \exp(i\omega t)$, where ω is the angular frequency of the laser, t is the time, and $i = \sqrt{-1}$ is the imaginary unit. Assume that the sample is infinitely thick so that $q/q_0 \to 0$ as $x \to \infty$:

   (a) Show that the one-dimensional GK equation can be written as

$$\tau_R \frac{\partial^2 q}{\partial t^2} + \frac{\partial q}{\partial t} = -\kappa \frac{\partial^2 q}{\partial x^2} + 3\ell^2 \frac{\partial^3 q}{\partial x^2 \partial t}, \tag{6.35}$$

   where $\kappa = k/(\rho_s c_s)$ is the bulk thermal diffusivity of the sample.

(b) By seeking a solution $q(x,t) = Q(x)\exp(i\omega t)$, show that the complex flux Q is given by

$$Q(x) = q_0 \exp\left(-i^{1/2}x/L_p\right), \tag{6.36}$$

where the complex thermal penetration depth is given by

$$L_p = \left(\frac{1 + 3i\ell^2\omega\kappa^{-1}}{1 + i\omega\tau_R}\right)^{1/2} L_0, \tag{6.37}$$

and $L_0 = (\kappa/\omega)^{1/2}$ is the classical thermal penetration depth.

(c) Determine the range of laser frequencies at which classical heat conduction is expected to occur. Deduce that a high-frequency laser is able to excite non-Fourier modes of heat conduction, even if the sample itself is not thin (i.e. thickness on the order of nanometres).

(d) Show that the change in surface temperature is given by

$$T(0,t) = \left(\frac{q_0}{i^{1/2}\rho c_s \omega L_p}\right)\exp(i\omega t). \tag{6.38}$$

Deduce that in the classical regime of thermal transport, the surface temperature will be out of phase with the laser with a phase shift of $\varphi = \arg(T(0,t)) = -\pi/4$. A phase shift that differs from $-\pi/4$ would therefore represent a signature of non-Fourier thermal transport. Show that the predicted phase shift in the presence of non-Fourier effects is given by

$$\varphi = -\frac{\pi}{4} + \frac{1}{2}\arctan(\omega\tau_R) - \frac{1}{2}\arctan\left(\frac{3\ell^2\omega}{\kappa}\right). \tag{6.39}$$

2. When observing the nondimensional equations (6.25) and (6.26), determine the relation that the nonclassical dimensionless parameters γ and η must satisfy for the system to be in a state of Fourier resonance.
3. Consider the one-dimensional nondimensional solidification problem given by (6.25)–(6.28). For $\beta \gg 1$, solidification is very slow compared to the heat conduction. Determine the leading order solution for the temperature and the size of the solid when $\gamma = \mathcal{O}(1)$ and $\eta = 0$.
4. Consider again the one-dimensional nondimensional solidification problem given by (6.25)–(6.28) with $\eta = 0$. Now suppose that γ is large and of the same order of magnitude as β. Therefore, let $\gamma = \beta\Gamma$ where $\Gamma = O(1)$ as $\beta \to \infty$.

(a) By rescaling time as $t = \beta\tau$, show that the leading order problem as $\beta \to \infty$ is given by

$$\frac{\partial q}{\partial x} = 0, \tag{6.40}$$

$$\Gamma \frac{\partial q}{\partial \tau} + q = -\frac{\partial T}{\partial x}, \tag{6.41}$$

with the Stefan condition

$$\frac{ds}{d\tau} = -q|_{x=s(t)}. \tag{6.42}$$

Deduce that the flux q must only be a function of time and the temperature is a linear function of space and hence given by $T = x/s(\tau) - 1$.

(b) Show that the solidification front satisfies a nonlinear second-order ODE given by

$$\Gamma \ddot{s} + \dot{s} = \frac{1}{s} \tag{6.43}$$

where dots denote first and second derivatives with respect to τ.

(c) Numerically solve (6.43) using the initial conditions $s(0) = 0$ and $\dot{s}(0) = 0$. Show that as $\tau \gg 1$, the solution approaches the classical solidification solution given by $s(\tau) = (2\tau)^{1/2}$. Hence, on long time-scales, the impact of non-Fourier conduction diminishes. Note: the initial condition for $\dot{s}$ can be determined using matched asymptotic expansions; see Ref. [24] for full details.

5. Solve the system (6.30) subject to $\bar{T}(0, \bar{t}) = -1$, $\bar{T} \to 0$ as $\bar{x} \to \infty$, $\bar{T}(\bar{x}, 0) = \bar{q}(\bar{x}, 0) = 0$.
6. A popular alternative to the Guyer-Krumhansl (GK) equation is the dual phase-lag (DPL) model:

$$\mathbf{q} + \tau_q \dot{\mathbf{q}} = -k\nabla T - k\tau_T \nabla \dot{T},$$

where the nonclassical parameters are two relaxation times τ_q and τ_T. In the context of the one-dimensional solidification process leading to Eqs. (6.20)–(6.23), show that both the DPL and GK models lead to the same type of third-order heat equation.

Chapter 7
Hints to Exercises

Chapter 1

1. $\tau = \rho c L^2/k$
   Substitute for the solid and liquid values: solid gold $\tau \approx 78.54$ s, liquid gold $\tau \approx 266$ s, ice $\tau \approx 8572$ s, and liquid water $\tau \approx 111,075$ s.
2. $\beta = \frac{\rho_s}{\rho_l}\frac{L_m}{c_l \Delta\theta}$.
3.
$$\frac{\partial \hat{T}}{\partial \hat{x}} = Nu(1 - \hat{T}) .$$

   The limit $Nu \to \infty$ gives $\hat{T} = 1$; $Nu \to 0$ gives $\hat{T}_{\hat{x}} = 0$.
   For the final condition, assume $\hat{T} \ll 1$ and so choose $\hat{T} = \varepsilon \hat{S}$, where $\varepsilon \ll 1$. This gives

$$\varepsilon \frac{\partial \hat{S}}{\partial \hat{x}} = Nu(1 - \varepsilon \hat{T}) .$$

   Choosing $\hat{x} = \varepsilon \hat{y}$

$$\frac{\partial \hat{S}}{\partial \hat{y}} = Nu(1 - \varepsilon \hat{T}) \approx Nu .$$

   In this example, the Neumann condition may be obtained when $\hat{T} \ll 1$ and close to the surface, such that $\hat{x} \ll 1$, or equivalently $T \ll T_a, \ \ x \ll L$.
4. For n-octadecane $\beta \approx 10$, gold $\beta \approx 50$ and ice $\beta \approx 16$.
   Naively considering $s_t \sim T_x/\beta$ suggests the larger the value of β, the lower the value of s_t (hence gold is slower), but the times are scaled differently for each material, so a faster nondimensional rate does not necessarily translate to a faster melting rate in reality.

T. G. Myers, *Theoretical and Practical Stefan Problems*, Interdisciplinary Applied Mathematics 63, https://doi.org/10.1007/978-3-032-04826-4_7

A better measure would be the melting time-scale, which suggests n-octadecane. $\tau = \rho_s L_m L^2/(k_s \Delta T) \approx 3900, 1.4 \times 10^5, 6.1 \times 10^5$ s for gold, water, and n-octadecane of 10 cm. Alternatively, the ratio of time-scales is independent of the size and temperature drop, i.e. $\tau_g/\tau_n \approx 6 \times 10^{-3}$ – gold melts much more rapidly than n-octadecane.

5. From Example 4, we see that n-octadecane is slow to melt, so that energy is slowly absorbed or released. Further, from Table 1.1, we see that the melting point $T_m = 301.3\,\text{K} \approx 28\,°\text{C}$ is a normal temperature in hot countries. Consequently, we may expect n-octadecane to solidify during the night when the temperature is below 28 °C and then melt during the day. The energy absorbed by the melting material is no longer available to heat the surroundings – resulting in a lower temperature rise. Effectively air conditioning without the need for electricity.
6. Denote $T_i = 450$ K, $T_m = 505$ K, $T_w = 600$ K; then

$$\hat{T} = \frac{T - T_m}{T_m - T_i} \in [-1, 0] \qquad \hat{\theta} = \frac{\theta - T_m}{T_w - T_m} \in [0, 1]$$

7. It's best to bring back the hat notation and then

$$\hat{T} = \frac{T - T_m}{T_i - T_m} \Rightarrow T = (T_i - T_m)\hat{T} + T_m$$
$$\hat{\theta} = \frac{\theta - T_m}{T_i - T_m} \Rightarrow \theta = (T_i - T_m)\hat{\theta} + T_m$$

The dimensional system is simply that specified by equations (1.1), (1.2). Subject to

$$T(0, t) = (T_i - T_m)\hat{T}_0 + T_m \ , \quad T(s, t) = \theta(s, t) = T_m \ ,$$
$$\theta|_{x \to \infty} = \theta(x, 0) = T_i \ .$$

8. Choose

$$\hat{c} = \frac{c - c_i}{c_i - c_p} \in [-1, 0] \qquad \hat{r}_p = \frac{r_p}{R} \qquad \hat{t} = \frac{t}{\tau}$$

where $\tau = R^2/D$ and R is an unspecified length-scale. The "Stefan number" is $1/(V_m \Delta c)$. The full governing equations are presented in Sect. 2.1.4.
9. See Sect. 3.2

Chapter 2

1. (a) Firstly, we need to swap the domains for θ, T, i.e. the liquid occupies $x \in [0, s)$ and the solid $x > s$. Melting is driven by heat flow through the liquid, so take the liquid time-scale and now $\beta = \rho_s L_m/(\rho_l c_l (T_m - T_i))$. They are

solved subject to

$$\theta(0,t) = T_0 > 0\,, \quad T(s,t) = \theta(s,t) = 0\,, \quad T|_{x\to\infty} = T(x,0) = -1\,.$$

The corresponding temperature profiles are

$$\theta = T_0\left(1 - \frac{\text{erf}\,\left(x/2\sqrt{t}\right)}{\text{erf}\,(\lambda)}\right), \quad T = -1 + \frac{\text{erfc}\,\left(x/(2\sqrt{\alpha t})\right)}{\text{erfc}(\lambda/\sqrt{\alpha})}\,,$$

where $\alpha = \alpha_s/\alpha_l$ and

$$\beta\lambda = \frac{1}{\sqrt{\pi}}\left[\frac{T_0 e^{-\lambda^2}}{\text{erf}\,(\lambda)} - \frac{k e^{-\lambda^2/\alpha}}{\sqrt{\alpha}\,\text{erfc}\,(\lambda/\sqrt{\alpha})}\right].$$

For uniqueness, one could consider each component – all separate functions are monotonic.

(b) This is a simple extension of the above analysis coupled to the analysis following (2.15).

2. The problem is described by

$$\frac{\partial\theta}{\partial t} = \frac{\partial^2\theta}{\partial x^2}\,, \quad 0 < x < s \qquad \beta\frac{ds}{dt} = -\left.\frac{\partial\theta}{\partial x}\right|_{x=s}\,,$$

subject to

$$\theta(0,t) = 1 \qquad \theta(s,t) = 0 \qquad s(0) = 0\,,$$

with a corresponding temperature of the form

$$\theta = A + B\,\frac{\text{erfc}\,\left(x/(2\sqrt{t})\right)}{\text{erfc}(\lambda)}$$

where the constants are chosen to satisfy the boundary conditions.
The melting process is faster than the corresponding two-phase model, since the solid now takes no energy away from the melting process:

3. Setting $\theta(x,t) = G(\eta(x,t),t)$, the liquid heat equation transforms to

$$s^2\frac{\partial G}{\partial t} - s\eta\frac{ds}{dt}\frac{\partial G}{\partial\eta} + s\left(1 - \frac{\rho_s}{\rho_l}\right)\frac{ds}{dt}\frac{\partial G}{\partial\eta} = \alpha\frac{\partial^2 G}{\partial\eta^2}\,,$$

The solid heat equation and Stefan condition are given by Eqs. (2.52, 2.53).

4.

$$T = T_0(1 - x/s)$$

In the pseudo-steady state, the zero flux condition on θ forces θ to be constant. The problem could be addressed by rescaling in a boundary layer near $x = s$ or applying a heat balance method on $x \geq s$ (but note heat balance methods with a finite domain require switching solution form when δ reaches the boundary; see [78]).

5. (a)

$$\theta = \frac{1}{1-R}\left(1 - \frac{R}{r}\right) \quad \Rightarrow \quad t = \left(\frac{1-R^2}{2}\right) - \left(\frac{1-R^3}{3}\right)$$

Full melting occurs when $R = 0$, $t = 1/6$.

(b) Now $u = (r - R)/(1 - R)$ and the leading order solution for R is the same.

6.

$$\theta = 1 - \frac{\ln r}{\ln R} \quad \Rightarrow \quad t = \frac{R^2 \ln R}{2} + \frac{1 - R^2}{4}$$

L'Hôpital's rule can show $\lim_{R \to 0} R^2 \ln R = 0$ and so the cylinder is melted when $t = 1/4$.

7. One anticipates the TIM to be most accurate for $\beta > 1$; for $\beta < 1$ as the value decreases, there will come a point where the perturbation wins.
8. The Stefan condition may now be written as

$$s s_t = \frac{2n(n+1)(1+a_1)}{2 + a_1(1-n)} \,.$$

Analysis and advantages of various approaches are discussed in [89].

9. For the implicit function, it is much simpler to choose a δ range, say $\delta \in (0.001, 1)$ and calculate t (if this range is not sufficient, then change it appropriately).
10. (a) Setting $\eta = x - s(t)$ leads to

$$\frac{\partial F}{\partial \eta} = s_t \frac{\partial G}{\partial \eta} \qquad - s_t \frac{\partial G}{\partial \eta} = 1 - G \,.$$

For this to be independent of time requires s_t constant; hence, we may write $s_t = v$, $s = vt + s_0$.

(b) The first equation determines $F = vG$ (after imposing $F = G = 0$ at $\eta = 0$). The condition at $\eta = -\infty$ determines $v = 1$. The second equation gives $G = 1 - \exp(\eta/v)$

(c) The system is now

$$(1 - vDa)\frac{\partial F}{\partial \eta} = v \frac{\partial G}{\partial \eta} \qquad - v \frac{\partial G}{\partial \eta} = 1 - G \,.$$

Integrating the first equation and applying the conditions $v = 1/(1 + Da)$ and $F = G$. You can figure out the rest.

11. (a) The standard diffusion time-scale $\tau_D = R^2/D$ and the growth time-scale $\tau_g = \rho_\beta R^2/(M_g(c_R - c_{eq})D)$.
The boundary condition becomes

$$\left.\frac{\partial \hat{c}}{\partial \hat{r}}\right|_{\hat{r}=\hat{s}} = \frac{kR}{D}\hat{c}(\hat{s},\hat{t}) = \frac{k\tau_D}{R}\hat{c}(\hat{s},\hat{t}),$$

which suggests defining $\tau_k = R/k$.

(b) Choosing $\tau = \tau_g$ and assuming $\tau_D/\tau_g = \varepsilon \ll 1$ leads to a leading order ODE, with solution of the form

$$\hat{c} = 1 + B - B/\hat{r}$$

where $B = B(s)$. The Stefan condition

$$\frac{d\hat{s}}{d\hat{t}} = -\frac{1}{1 + T_m\hat{s} - T_m\hat{s}^2}$$

where $T_m = Rk/D$. This leads to

$$T_m\frac{\hat{s}^3 - 1}{3} - T_m\frac{\hat{s}^2 - 1}{2} - (\hat{s} - 1) = \hat{t}.$$

The core is all used up when $\hat{s} = 0$,

$$\hat{t}_f = 1 + \frac{T_m}{6}.$$

(c) The *diffusion controlled* model is obtained in the limit $T_m \gg 1$. After dividing through by T_m,

$$\frac{\hat{s}^3 - 1}{3} - \frac{\hat{s}^2 - 1}{2} = \hat{t}/T_m.$$

The term $(\hat{s}-1)/T_m$ has been neglected under the assumption that it is small; however, $\hat{t}/T_m$ is retained. This is only valid for $\hat{t} = \mathcal{O}(T_m)$. Consequently, this solution can only be applied (and compared with data) for large times.

Chapter 3

1. No-slip implies $u(0) = u(H) = 0$ and we may write

$$u = \frac{p_x}{2\mu} z(z - H) \quad \Rightarrow \quad Q = \int_0^H u \, dz = -\frac{p_x H^3}{12\mu}$$

$$\Rightarrow \quad p = p_a + \frac{12\mu Q}{H^3}(L - x)$$

2.

$$p = p_a + \frac{6\mu Q}{R^3}(L - x)$$

3. With $Q = A = 0$, we have $p_x = 0$. Equation (3.27) indicates $Bh_x - Ch_{xxx} = 0$ with solution

$$h = c_0 + c_1 e^{\sqrt{B/C}\,x} - c_2 e^{-\sqrt{B/C}\,x}.$$

The constants come from the boundary conditions.
When $B/C \gg 1$, the surface is mostly flat – the solution appears constant at leading order and surface tension must be retained to allow $h \to 0$. When surface tension dominates, the surface takes a quadratic form, i.e. the droplet is curved. Try dropping water on a table – large drops are mainly flat, and small ones are highly curved.

4.

$$p|_{x=1} - p|_{x=0} = -\frac{3Q}{2(1 - \sqrt{2t})^3}$$

Obviously as $t \to 1/2$, the pressure drop tends to infinity. This indicates that as the channel closes, it becomes increasingly difficult to force fluid through.

5.

$$s^2(1 - 2\ln s) = 1 - 4t$$

From L'Hôpital, we find $\lim_{s \to 0} s^2 \ln s = 0$ and so $t_{cl} = 1/4$.

6. Firstly, from incompressibility,

$$u_x + w_z = 0 \quad \Rightarrow \quad w(x, h) - w(x, 0) = -\int_0^h u_x \, dx$$

From this, we obtain a relation between h_t and p_{xx}. This may be integrated, subject to $p(x = \pm 1) = 0$ to determine p, and finally evaluating the force integral gives an equation for h_t.

7. The neglected terms are $\varepsilon^2 T_{xx}, \alpha\varepsilon^2\theta_{xx}$.

8. In the solid there is no x or y variation and the height scale is not small so there is no dominance of the diffusion term. Start with the Laplacian of the liquid temperature:

$$\left(\frac{\partial^2\theta}{\partial x^2}+\frac{\partial^2\theta}{\partial y^2}+\frac{\partial^2\theta}{\partial z^2}\right)\to\frac{\Delta\theta}{H^2}\left(\frac{H^2}{L^2}\frac{\partial^2\theta}{\partial x^2}+\frac{H^2}{B^2}\frac{\partial^2\theta}{\partial y^2}+\frac{\partial^2\theta}{\partial z^2}\right)$$

So the first two terms may be neglected. The advection term

$$\left(u\frac{\partial\theta}{\partial x}+v\frac{\partial\theta}{\partial y}+w\frac{\partial\theta}{\partial z}\right)\to\frac{U\Delta\theta}{L}\left(u\frac{\partial\theta}{\partial x}+v\frac{\partial\theta}{\partial y}+w\frac{\partial\theta}{\partial z}\right)$$

The presence of the $1/H^2$ term in the diffusion allows the neglect of advection and the time derivative.

9. With the eigenfunction $F = A_n\sin(\lambda_n(z-H)) + B_n\cos(\lambda_n(z-H))$, the condition at $z = H(t)$ determines $A_n = 0$, the condition at $z = h(t)$ gives $\lambda_n = (2n-1)\pi/(2(h(t)-H(t))$, and the "constant" of separation is a function of time.

10. (a)

$$T = 1 + (T_0 - 1)\mathrm{erf}\left(z\sqrt{\frac{A_1}{4t}}\right).$$

Melting first occurs at $z = 0$ and, since $\mathrm{erf}(0) = 0$, then $T(0, t) = 1$ for all time – melting occurs immediately (its infinitely high heat transfer!).

(b)

$$T = T_0 + (1 - T_0)\left(1-\frac{z}{\delta}\right)^n, \qquad \delta = \sqrt{\frac{2n(n+1)t}{A_1(1+nT_0)}}$$

and for the TIM $n = 2.233$. This is only valid provided $\delta \le 1$.

11.

$$\int_0^1 \Phi_n\,dY = \frac{\lambda^2 F_n}{\omega_n^3}\left[\frac{(2\lambda-\omega_n)-(2\lambda+\omega_n)e^{-\omega_n/\lambda}}{1+e^{-\omega_n/\lambda}}\right].$$

Hence,

$$\begin{aligned}1-h_m(t) &= \sum_{n=1}^{\infty}\int_0^1\int_0^1 \Phi_n(Y,t)\sin(\omega_n X)\,dY\,dX\\ &= \sum_{n=1}^{\infty}\frac{8f(t)(1-(-1)^n)^2}{\omega_n^5}\left[\frac{(2\lambda-\omega_n)-(2\lambda+\omega_n)e^{-\omega_n/\lambda}}{1+e^{-\omega_n/\lambda}}\right].\end{aligned}$$

Now, assume $e^{-\omega_n/\lambda}$ is negligible and take only the $n = 1$ term:

$$1 - h_m(t) = \frac{32(2\lambda - \pi)}{\pi^5} f(t) = \Lambda f(t)\,.$$

The first equation of (3.115) remains unchanged; the second follows from $\lambda \to 0$ in the above expression $1 - h_m(t) = -32/\pi^4 f(t)$.

12. Integrate

$$\frac{s}{1 - As}\frac{\partial s}{\partial t} = 1$$

and then

$$h = \frac{U\tau}{\rho_l L}\dot{m}t - \frac{\rho_s}{\rho_l}s\,.$$

13. From the definition of the enthalpy, the energy flux boundary condition may be written as

$$-k_l\frac{\partial\theta}{\partial z} = E_0 + E_1\theta + (c_l(\theta - T_m) + L_m - \mathrm{h}_{inc})|_{z=h_T} = E_0' + E_1'\theta\big|_{z=h_T}\,, \tag{7.2}$$

where E_0', E_1' contain the energy terms from the enthalpy. The PSS approximation gives

$$\theta = T_s - \frac{E_0' + E_1' T_s}{k_l + E_1' h_T} z \tag{7.3}$$

From setting $\theta(h_T, t) = T_m$, we obtain an expression for h_T in terms of the model parameters. From the mass balance

$$\rho\frac{dh_T}{dt} = \dot{m}_{inc} - \dot{m}_{evap} \quad \Rightarrow \quad \rho h_T = (\dot{m}_{inc} - \dot{m}_{evap})t\,, \tag{7.4}$$

we may relate h_T and t.

Chapter 4

1.

$$\frac{1}{2}\rho_s v_l^2 = \left(1 - \frac{\rho_s}{\rho_l}\right) p_l$$

Since the left-hand side is nonnegative, when $\rho_s > \rho_l$, the fluid pressure $p_l < 0$. This indicates a negative pressure (suction), which acts to move the fluid backwards towards the surface, which is effectively moving away from the fluid as it contracts upon phase change.

2. The relative size comes through comparison of

$$\frac{1}{2}\left(1-\left(\frac{\rho_s}{\rho_l}\right)^2\right)s_t^2=\frac{1}{2}\left(1-\left(\frac{\rho_s}{\rho_l}\right)^2\right)\frac{\lambda^2}{t}, \quad (c_l-c_s)(T_I-T_m), \quad L_m .$$

They are therefore comparable at

$$t_1=\frac{1}{2}\left(1-\left(\frac{\rho_s}{\rho_l}\right)^2\right)\frac{\lambda^2}{L_m}, \quad t_2=\frac{1}{2}\left(1-\left(\frac{\rho_s}{\rho_l}\right)^2\right)\frac{\lambda^2}{(c_l-c_s)(T_I-T_m)} .$$

Giving $t_1 \approx 4.7\times 10^{-9}$s, $t_2 \approx 7.4\times 10^{-8}$s.
We may believe that it is sensible to neglect kinetic energy when latent heat is approximately ten times the kinetic energy, so $t \approx 4.7\times 10^{-8}$ s. Consequently, at least for standard models, for the freezing of water, kinetic energy plays a insignificant role, but when we move on to nanoscale freezing (working on the picosecond scale), kinetic energy may dominate for much of the process.

3. Mass balance:

$$\frac{1}{r}\frac{\partial}{\partial r}(rv_l)=0 \quad\Rightarrow\quad v_l(r,t)=A(t)/r .$$

Jump condition at $r=R(t)$

$$(\rho_l-\rho_s)\frac{dR}{dt}=\rho_l v_l(R,t) \quad\Rightarrow\quad v_l(r,t)=\left(1-\frac{\rho_s}{\rho_l}\right)\frac{R}{r}\frac{dR}{dt} .$$

Then

$$v_l(R_b,t)=\mathrm{d}R_b/\mathrm{d}t \quad\Rightarrow\quad R_b^2=\frac{\rho_s}{\rho_l}R_0^2+\left(1-\frac{\rho_s}{\rho_l}\right)R^2 ,$$

where we have imposed $R_b(0)=R(0)=R_0$.

4. (a) Temperature profiles

$$T=T_m(t) \qquad \theta=(T_m-1)\frac{\ln r}{\ln R}+1$$

(b) Hence,

$$\begin{aligned}\frac{dR}{dt}&=k\frac{\partial T}{\partial r}-\frac{\partial\theta}{\partial r}=-\frac{\partial\theta}{\partial r}\\&=\frac{1-T_m}{R\ln R}=\frac{2R+\Gamma}{2R^2\ln R}\end{aligned}$$

Obviously, this is infinite at $R = 0, 1$. For $R = 1$, there is an instantaneous jump in temperature due to the conditions $\theta(1, t) = 1, \theta(R, t) = -\Gamma/(2R)$ (and no, $-\Gamma/(2R) \neq 1$). As $R \to 0$, the solid surface area also $\to 0$, consequently the energy required to cause melting $\to 0$.

(c) When $\Gamma = 0$, we find

$$t = \frac{R^2 \ln R}{2} - \frac{R^2 - 1}{4}.$$

5. (a) Since the solid is a better conductor, the cooling through the solid will be more rapid than the equivalent heating through a liquid. Hence, we expect the solidification to be faster.

(b)

$$\frac{dR}{dt} = \frac{k(2R + \Gamma)}{2R^2 \ln R}$$

The factor $k \gg 1$ verifies that solidification will be more rapid than melting.

6. Simply put the values into the energy equation (4.47).
7. (a)

$$\theta = -1 + \frac{\text{erfc}\left(x/(2\sqrt{t})\right)}{\text{erfc}(\lambda)},$$

As $\beta \to 1, \lambda \to \infty$ and the front moves with infinite velocity.

(b)

$$E = \int_s^\infty (\theta + \beta)\, dx$$

$$\frac{dE}{dt} = \int_s^\infty \theta_t\, dx - (\theta(s, t) + \beta)s_t = -\theta_x(s, t) - \beta s_t = 0$$

The system conserves energy.

(c)

$$\frac{dE}{dt} = -\theta(s, t)s_t = s_t^2 \neq 0$$

The system does not conserve energy.

8. (a) The simplest way to achieve this is to leave the derivatives inside the brackets in their original form, e.g. $\partial(\rho_s u_s)/\partial t$. In this way, Eq. (4.1d) may be invoked to replace the time derivative with the space derivative, which may then be immediately integrated. Don't forget the derivatives of the integral limits.

(b) Equation (4.13) shows that the terms evaluated at $r = R$ cancel out (they define both sides of the Stefan condition).

(c) Taking the time derivative of the energy balance and substituting for the previous energy derivatives leads to

$$-q_l = k_l \frac{\partial \theta}{\partial r}\bigg|_{R_b} = \mathscr{H}_{la}(T_a - \theta(R_b, t)) .$$

This is, of course, the standard Newton cooling condition.

Chapter 5

1.

$$\theta = -1 + (1 + T_I)\left(\frac{\delta - x}{\delta - s}\right)^n$$

The heat balance integral produces

$$\frac{d}{dt}\left[\frac{(T_I - n)(\delta - s)}{n+1}\right] + \delta_t + T_I s_t = \frac{n(1 + T_I)}{\delta - s}$$

This must be solved together with the Stefan condition.
If your code crashed at the start

$$s = t - (2/3)\sqrt{\beta^2(n+1)t^3/n}$$

is the small time solution, use this for some $t = \varepsilon \ll 1$, as the initial condition. Full details may be found in [52].

2. Plot graphs of equations (5.57), (5.67).
3. For a sphere $\kappa = 2/R$, appropriate parameter values are provided in Table 1.1 For gold, the graph becomes multivalued for $T_I \approx 340$ K, $R \approx 0.95$ nm. For water, the graph becomes multivalued for $T_I \approx 155$ K, $R \approx 0.82$ nm.

4. (a)

$$c = \frac{A}{r} + B \qquad \frac{dr_p}{dt} = -\frac{V_m D A}{r_p^2} \qquad A = \frac{(S - c_b)kr_p^2(r_p + \delta)}{\delta k r_p + D(r_p + \delta)}$$

$$B = c_b - \frac{A}{r_p + \delta}$$

(b)

$$c = c_b + \frac{(c_b - c_s)kr_p^2}{(D + kr_p)\delta}\left(1 - \frac{\delta}{r}\right), \qquad \frac{dr_p}{dt} = -\frac{V_m D(S - c_b)k}{(D + kr_p)}$$

(c)

$$a^3 = V_m(c_0 - S_{eq}), \qquad b^3 = \frac{4\pi N_0}{3}$$

5. The simplest method is to plot the two solutions. Also try expanding the expression for r_p for t close to t_s.
6. The solid temperature is unchanged, and the liquid temperature takes the form

$$\theta = \frac{A}{r} + B \qquad A = Nu\left[1 + \frac{\Gamma}{R}\right]\left[\frac{Nu}{R_b} - \frac{Nu}{R} - \frac{1}{R_b^2}\right]^{-1}, \qquad B = -\frac{\Gamma}{R} - \frac{A}{R}$$

In the limit, $Nu \to \infty$ and

$$A \to -\left[1 + \frac{\Gamma}{R}\right]\frac{RR_b}{R_b - R}$$

From this, the expression for θ_0 in (5.94) is retrieved; hence, the plots must also coincide as Nu is increased
7. The analysis basically follows the leading order of Sect. 5.4, leading to

$$T = -\frac{\Gamma}{R} \qquad \theta = 1 - \left(1 + \frac{\Gamma}{R}\right)\frac{\ln r}{\ln R}$$

The Stefan condition then has

$$\frac{dR}{dt} = -\left.\frac{\partial\theta}{\partial r}\right|_{r=R} = \left(1 + \frac{\Gamma}{R}\right)\frac{1}{R\ln R}$$

When $\rho \neq 1$, we follow the analysis of Sect. 4.3.2 to find

$$v_l = (1-\rho)\frac{R}{r}\frac{dR}{dt} \qquad R_b^2 = \rho + (1-\rho)R^2$$

and

$$T = -\frac{\Gamma}{R} \qquad \theta = 1 - \left(1 + \frac{\Gamma}{R}\right)\frac{\ln(r/R_b)}{\ln(R/R_b)}$$

and

$$\frac{dR}{dt} = \left(1 + \frac{\Gamma}{R}\right)\frac{1}{R\ln(R/R_b)}$$

Chapter 6

1. (c): $\omega \ll \tau_R^{-1}$ and $\omega \ll 3\ell^2/\kappa$.
2. $\gamma = \eta^2$ or $\zeta = 1$.
3. The resulting hyperbolic heat equation is linear and solidification is neglected at leading order. To leading order, $s = 1$. The solution for the temperature may be obtained using separation of variables.
4. (a) Differentiate the nondimensional MC equation (6.41) with respect to x and then apply the boundary conditions. (b) First, show that the flux $q(t)$ satisfies the ODE given by $\Gamma\dot{q} + q = -1/s$. Then, differentiate the Stefan condition with respect to τ and eliminate $\dot{q}$ and q to produce the ODE for s.
5. Use the similarity variable $\eta = x/\sqrt{t}$; the answer is given by (6.31).
6. For both models, the associated one-dimensional heat equation is of the form

$$a\frac{\partial T}{\partial t} + b\frac{\partial^2 T}{\partial t^2} = c\frac{\partial^2 T}{\partial x^2} + d\frac{\partial^3 T}{\partial x^2 \partial t}.$$

Bibliography

1. M. Abramowitz, I.A. Stegun, *Handbook of Mathematical Functions with Formulas, Graphs, and Mathematical Tables*, 9th edn. (Dover, New York, 1972)
2. D.J. Acheson, *Elementary Fluid Dynamics*, 12th edn. (Oxford University Press, Oxford, 2005)
3. M. Aguareles, E. Barrabés, T.G. Myers, A. Valverde, Mathematical analysis of a Sips-based model for column adsorption. Physica D: Nonlinear Phenomena **448**, 133690 (2023)
4. V. Alexiades, A.D. Solomon, *Mathematical Modelling of Freezing and Melting Processes*, 1st edn. (Hemisphere Publishing Corporation, New York, 1993)
5. F.X. Alvarez, D. Jou, Size and frequency dependence of effective thermal conductivity. J. Appl. Phys. **103**(9), 094321 (2008)
6. J.G. Andrews, D.R. Atthey, On the motion of an intensely heated evaporating boundary. IMA J. Appl. Math. **15**(1), 59–72 (1975)
7. M.F. Ashby, D.R.H. Jones, *Engineering Materials 2: An Introduction to Microstructures, Processing and Design*, 3rd edn. (Butterworth-Heinemann, Oxford, 2006)
8. J. Back, *Stefan Problems for Melting Nanoscaled Particles* (Queensland University of Technology, Brisbane, 2014)
9. J.M. Back, S.W. McCue, T.J. Moroney, Including nonequilibrium interface kinetics in a continuum model for melting nanoscaled particles. Sci. Rep. **4**, 7066 (2014)
10. J.M. Back, S.W. McCue, M.H-N. Hsieh, T.J. Moroney, The effect of surface tension and kinetic undercooling on a radially-symmetric melting problem. Appl. Math. Comput. **229**, 41–52 (2014)
11. S.I. Barry, J. Caunce, Exact and numerical solutions to a Stefan problem with two moving boundaries. Appl. Math. Model. **32**(1), 83–98 (2008)
12. N.G Bastús, J. Comenge, V. Puntes, Kinetically controlled seeded growth synthesis of citrate-stabilized gold nanoparticles of up to 200 nm: size focusing versus Ostwald ripening. Langmuir **27**(17), 11098–11105 (2011)
13. A. Bejan, *Convection Heat Transfer*, 3rd edn. (Wiley, Hoboken, 2004)
14. C.M. Bender, S.A. Orszag, *Advanced Mathematical Methods for Scientists and Engineers I: Asymptotic Methods and Perturbation Theory* (Springer, Berlin, 1999)
15. R.B. Bird, W.E. Stewart, E.N. Lightfoot, *Transport Phenomena*, 2nd edn. (Wiley, Hoboken, 2007)
16. T.W. Brakel, J.P.F. Charpin, T.G. Myers, One-dimensional ice growth due to incoming supercooled droplets impacting on a thin conducting substrate. Int. J. Heat Mass Transf. **50**, 1694–1705 (2007)

T. G. Myers, *Theoretical and Practical Stefan Problems*, Interdisciplinary Applied Mathematics 63, https://doi.org/10.1007/978-3-032-04826-4

17. A. Bucknell, M. McGilvray, D. Gillespie, X. Yang, et al., ICICLE: A model for glaciated & mixed phase icing for application to aircraft engines. SAE Technical Paper, no. 2019–01–1969 (2019)
18. A. Bucknell, M. McGilvray, D.R.H. Gillespie, G. Jones, B. Collier, A thermodynamic model for ice crystal accretion in aircraft engines: EMM-C. Int. J. Heat Mass Transf. **174**, 121270 (2021)
19. P. Buffat, J.P. Borel, Size effect on the melting temperature of gold particles. Phys. Rev. A **13**(6), 2287–2297 (1976)
20. J. Buongiorno et al., A benchmark study on the thermal conductivity of nanofluids. J. Appl. Phys. **106**(9), 094312 (2009)
21. M.H. Buschmann et al., Correct interpretation of nanofluid convective heat transfer. Int. J. Thermal Sci. **129**, 504–531 (2018)
22. C. Cagran, B. Wilthan, G. Pottlacher, Enthalpy, heat of fusion and specific electrical resistivity of pure silver, pure copper and the binary Ag–28Cu alloy. Thermochimica Acta **445**(2), 104–110 (2006)
23. J. Caldwell, Y.Y. Kwan, Numerical methods for one-dimensional Stefan problems. Commun. Numer. Methods Eng. **20**, 535–545 (2004)
24. J. Caldwell, Y.Y. Kwan, A brief review of several numerical methods for one-dimensional Stefan problems. Thermal Sci. **13**(2), 61–72 (2009)
25. M. Calvo-Schwarzwälder, Non-local effects and size-dependent properties in Stefan problems with Newton cooling. Appl. Math. Model. **76**, 513–525 (2019)
26. M. Calvo-Schwarzwälder, M.G. Hennessy, P. Torres, T.G. Myers, F.X. Alvarez, A slip-based model for the size-dependent effective thermal conductivity of nanowires. Int. Commun. Heat Mass Transf. **91**, 57–63 (2018)
27. M. Calvo-Schwarzwälder, M.G. Hennessy, T.G. Myers, The one-dimensional Stefan problem with non-Fourier heat conduction. Int. J. Thermal Sci. **150**, 106210 (2020)
28. Y. Cao, S. Hou, Extension to the Myers model for calculation of three-dimensional glaze icing. J. Aircraft **53**(1), 1–11 (2015)
29. H.S. Carslaw, J.C. Jaeger, *Conduction of Heat in Solids*, 2nd edn. (Oxford Scientific Publications, Oxford, 1959)
30. C. Cattaneo, A form of heat conduction equation which eliminates the paradox of instantaneous propagation. Compte Rendus **247**(4), 431–433 (1958)
31. I.E. Chang, Navier-Stokes solutions at large distances from a finite body. J. Math. Mech. **10**, 811–876 (1961)
32. C. Charach, B. Zaltzman, Planar solidification from an undercooled melt: Asymptotic solutions to a continuum model with interfacial kinetics. Phys. Rev. E **47**(2), 1230–1234 (1993)
33. C. Charach, B. Zaltzman, Analytic model for planar growth of a solid germ from an undercooled melt. Phys. Rev. E **49**(5), 4322–4327 (1994)
34. Z. Chen, J. Wang, S. Qian, H.H. Bau, Thermally-actuated, phase change flow control for microfluidic systems. Lab Chip **5**, 1277–1285 (2005)
35. J. Crank, *The Mathematics of Diffusion*, 2nd edn. (Oxford University Press, Oxford, 1975).
36. J. Crank, *Free and Moving Boundary Problems*, 1st edn. (Oxford University Press, Oxford, 1984)
37. R.V. Craster, O.K. Matar, Dynamics and stability of thin liquid films. Rev. Mod. Phys. **81**, 1131–1198 (2009)
38. V. Cregan, J. Williams, T.G. Myers, Contact melting of a rectangular block with temperature-dependent properties. Int. J. Thermal Sci. **150**, 106218 (2020)
39. T.B. David, Y. Lereah, G. Deutscher, R. Kofman, P. Cheyssac, Solid-liquid transition in ultra-fine lead particles. Philos. Mag. A **71**(5), 1135–1143 (1995)
40. S.H. Davis, *Theory of Solidification* (Cambridge University Press, Cambridge, 2001)
41. P.G. Debenedetti, F. Stillinger, Supercooled liquids and the glass transition. Nature **410**, 259–267 (2001)

42. P.G. Debenedetti, *Metastable Liquids, Concepts and Principles* (Princeton University Press, Princeton, 1996)
43. Y. Dong, B.-Y. Cao, Z.-Y. Guo, Size dependent thermal conductivity of Si nanosystems based on phonon gas dynamics. Physica E **56**, 256–262 (2014)
44. U. Düring, J.H. Bilgram, W. Känzing, Properties of the solid-liquid interface of growing salol crystals: A dynamic light scattering investigation. Phys. Rev. A **30**(2), 946–959 (1984)
45. M. El-Hachem, Mathematical Models of Biological Invasion. PhD Thesis, Queensland University of Technology, 2022
46. J.D. Evans, J.R. King, Asymptotic results for the Stefan problem with kinetic undercooling. Quart. J. Mech. Appl. Math. **53**, 449–473 (2000)
47. C. Fanelli, V. Cregan, F. Font, T.G. Myers, Modelling nanocrystal growth via the precipitation method. Int. J. Heat Mass Transf. **165**, 120643 (2021)
48. A.V. Fedorov, A.V. Shulgin, Mathematical modeling of melting of nano-sized metal particles. Combust. Explos. Shock Waves **47**(2), 147–152 (2011)
49. B.J. Florio, T.G. Myers, The melting and solidification of nanowires. J. Nanopart. Res. **18**, 168 (2016).
50. F. Font, A one-phase Stefan problem with size-dependent thermal conductivity. Appl. Math. Model. **63**, 172–178 (2018)
51. F. Font, T.G. Myers, Spherically symmetric nanoparticle melting with a variable phase change temperature. J. Nanopart. Res. **15**(12), 1–13 (2013)
52. F. Font, S.L. Mitchell, T.G. Myers, One-dimensional solidification of supercooled melts. Int. J. Heat Mass Transf. **62**, 411–421 (2013)
53. F. Font, T.G. Myers, S.L. Mitchell, Mathematical model for nanoparticle melting with density change. Microfluid. Nanofluidics **18**, 233–243 (2015)
54. A.C. Fowler, *Mathematical Models in the Applied Sciences*, 1st edn. (Cambridge University Press, Cambridge, 1997)
55. G.R. Fulford, P. Broadbridge, *Industrial Mathematics: Case Studies in the Diffusion of Heat and Matter*. Australian Mathematical Society Lecture Series, 1st edn. (Cambridge University Press, Cambridge, 2001)
56. T.R. Goodman, The heat-balance integral and its application to problems involving a change of phase. Trans. ASME **80**, 335–342 (1958)
57. D. Groulx, M. Lacroix, Study of close contact melting of ice from a sliding heated flat plate. Int. J. Heat Mass Trans. **49**, 4407–4416 (2006)
58. Y. Guo, M. Wang, Phonon hydrodynamics for nanoscale heat transport at ordinary temperatures. Phys. Rev. B **97**, 035421 (2018)
59. S.C. Gupta, *The Classical Stefan Problem*, 1st edn. (Elsevier, Amsterdam, 2003)
60. Y. Ha, Y-J Kim, T.G. Myers, On the numerical solution of a driven thin film equation. J. Comput. Phys. **227**, 7246–7263 (2008)
61. H.E. Huppert, M.G. Worster, Dynamic solidification of a binary melt. Nature **314**(25), 703–707 (1985)
62. M.G. Hennessy, M. Calvo-Schwarzwälder, T.G. Myers, Asymptotic analysis of the Guyer-Krumhansl-Stefan model for nanoscale solidification. Appl. Math. Model. **61**, 1–17 (2018)
63. M.G. Hennessy, M. Calvo-Schwarzwälder, T.G. Myers, Modelling ultra-fast nanoparticle melting with the Maxwell-Cattaneo equation. Appl. Math. Model. **69**, 201–222 (2019)
64. J.M. Hill, *One-Dimensional Stefan Problems: An Introduction*, 1st edn. (Longman Scientific & Technical, Harlow, 1987)
65. E.J. Hinch, *Perturbation Methods* (Cambridge University Press, Cambridge, 2000)
66. S. Howison, *Practical Applied Mathematics: Modelling, Analysis, Approximation* (Cambridge University Press, Cambridge, 2005)
67. D. Hughes, Safety group highlights CFIT risk for regionals. Aviation Week Space Technol. **140**, 46 (1994)
68. E. Javierre, C. Vuik, F.J. Vermolen, S. van der Zwaag, A comparison of numerical models for one-dimensional stefan problems. J. Comput. Appl. Math. **192**(2), 445–459 (2006)

69. D. Jou, G. Lebon, J. Casas-Vázquez, *Extended Irreversible Thermodynamics* (Springer, Berlin, 2010)
70. R. Kofman, P. Cheyssac, Y. Lereah, A. Stella, Melting of clusters approaching 0D. Eur. Phys. J. D **9**(1–4), 441–444 (1999)
71. V.K. La Mer, Nucleation in phase transitions. Ind. Eng. Chem. **44**(6), 1270–1277 (1952)
72. P.A. Lagerstrom, R.G. Casten, Basic concepts underlying singular perturbation techniques. SIAM Rev. **14**(1), 63–120 (1972)
73. O. Levenspiel, *Chemical Reaction Engineering*, 3rd edn. (Wiley, Hoboken, 1999)
74. D. Li, Y. Wu, P. Kim, L. Shi, P. Yang, A. Majumdar, Thermal conductivity of individual silicon nanowires. Appl. Phys. Lett. **83**(14), 2934–2936 (2003)
75. J.R. Lister, P.J. Dellar, Solidification of pressure-driven flow in a finite rigid channel with application to volcanic eruptions. J. Fluid Mech. **323**, 267–283 (1996)
76. H. Löwen, J. Bechhoefer, L.S. Tuckerman, Crystal growth at long times: critical behaviour at the crossover from diffusion to kinetics-limited regimes. Phys. Rev. A **45**(4), 2399–2415 (1992)
77. Y. Ma, Size-dependent thermal conductivity in nanosystems based on non-Fourier heat transfer. Appl. Phys. Lett. **101**, 211905 (2012)
78. M.M. MacDevette, T.G. Myers, Contact melting of a three-dimensional phase change material on a flat substrate. Int. J. Heat Mass Transf. **55**(23), 6798–6807 (2012)
79. M.M. MacDevette, T.G. Myers, Nanofluids: an innovative phase change material for cold storage systems? Int. J. Heat Mass Transf. **92**, 550–557 (2016)
80. M.M. MacDevette, T.G. Myers, B.R. Wetton, Boundary layer analysis and heat transfer of a nanofluid. Microfl. Nanofl. **17**, 401–412 (2014)
81. Y.D. Mao, M.T. Xu, Non-fourier heat conduction in a thin gold film heated by an ultra-fast-laser. Sci. China Technol. Sci. **58**, 638–649 (2015)
82. S.W McCue, B. Wu, J.M. Hill, Micro/nanoparticle melting with spherical symmetry and surface tension. IMA J. Appl. Math. **74**, 439–457 (2009)
83. J. Mei, J.W. Davenport, Molecular-dynamics study of self-diffusion in liquid transition metals. Phys. Rev. B **42**(15), 9682–9684 (1990)
84. M. Merder, Solidification fronts with unusual long-time behaviour. Phys. Rev. A **45**(4), 2158–2160 (1992)
85. B.L. Messinger, Equilibrium temperature of an unheated icing surface as a function of air speed. J. Aeronaut. Sci. **20**(1), 29–42 (1953)
86. S.L. Mitchell, Applying the combined integral method to one-dimensional ablation. Appl. Math. Model. **36**, 127–138 (2012)
87. S.L. Mitchell, T.G. Myers, Approximate solution methods for one-dimensional solidification from an incoming fluid. Appl. Math. Comput. **202**, 311–326 (2008)
88. S.L. Mitchell, T.G. Myers, Heat balance integral method for one-dimensional finite ablation. J. Thermophys. Heat Trans. **22**(3), 508–514 (2008)
89. S.L. Mitchell, T.G. Myers, Application of standard and refined heat balance integral methods to one-dimensional Stefan problems. SIAM Rev. **52**(1), 57–86 (2010)
90. S.L. Mitchell, T.G. Myers, Improving the accuracy of heat balance integral methods applied to thermal problems with time dependent boundary conditions. Int. J. Heat Mass Transf. **53**, 3540–3551 (2010)
91. S.L. Mitchell, T.G. Myers, Application of heat balance integral methods to one-dimensional phase change problems. Int. J. Diff. Equ. **2012**, 102879 (2012)
92. S.L. Mitchell, M. Vynnycky, Finite-difference methods with increased accuracy and correct initialization for one-dimensional Stefan problems. Appl. Math. Comput. **215**, 1609–1621 (2009)
93. M.K. Moallemi, B.W. Webb, R. Viskanta, An experimental and analytical study of close-contact melting. J. Heat Transf. **4**, 894–899 (1986)
94. E.B. Moore, V. Molinero, Structural transformation in supercooled water controls the crystallization rate of ice. Nature **479**, 506–508 (2011)

95. W.D. Murray, F. Landis, Numerical and machine solutions of transient heat-conduction problems involving melting or freezing: Part I — Method of analysis and sample solutions. ASME J. Heat Transf. **81**(2), 106–112 (1959)
96. T.G. Myers, Thin films with high surface tension. SIAM Rev. **40**(3), 441–462 (1998)
97. T.G. Myers, Extension to the Messinger model for aircraft icing. AIAA J. **39**(2), 211–218 (2001)
98. T.G. Myers, Optimizing the exponent in the heat balance and refined integral methods. Int. Commun. Heat Mass Transf. **36**, 143–147 (2009)
99. T.G. Myers, An approximate solution method for boundary layer flow of a power law fluid over a flat plate. Int. J. Heat Mass Transf. **53**, 2337–2346 (2010)
100. T.G. Myers, Optimal exponent heat balance and refined integral methods applied to Stefan problems. Int. J. Heat Mass Transf. **53**, 1119–1127 (2010)
101. T.G. Myers, Why are slip lengths so large in carbon nanotubes? Microfluid. Nanofluidics **10**(5), 1141–1145 (2011)
102. T.G. Myers, Mathematical modelling of phase change at the nanoscale. Int. Commun. Heat Mass Transf. **76**, 59–62 (2016)
103. T.G. Myers, Is it time to move on from the Bohart-Adams model for column adsorption? Int. Commun. Heat Mass Transf. **159**, 108062 (2024)
104. T.G. Myers, J.P.F. Charpin, A mathematical model for atmospheric ice accretion and water flow on a cold surface. Int. J. Heat Mass Transf. **47**(25), 5483–5500 (2004)
105. T.G. Myers, C. Fanelli, On the incorrect use and interpretation of the model for colloidal, spherical crystal growth. J. Colloid Interface Sci. **536**, 98–104 (2019)
106. T.G. Myers, F. Font, On the one-phase reduction of the Stefan problem with a variable phase change temperature. Int. Commun. Heat Mass Transf. **61**, 37–41 (2015)
107. T.G. Myers, F. Font, Mass transfer from a fluid flowing through a porous media. Int. J. Heat Mass Transf. **163**, 120374 (2020)
108. T.G. Myers, D.W. Hammond, Ice and water film growth from incoming supercooled droplets. Int. J. Heat Mass Transf. **42**, 2233–2242 (1999)
109. T.G. Myers, J. Low, An approximate mathematical model for solidification of a flowing liquid in a microchannel. Microfluid. Nanofluidics **11**, 417–428 (2011)
110. T.G. Myers, J. Low, Modelling the solidification of a power-law fluid flowing through a narrow pipe. Int. J. Thermal Sci. **70**, 127–131 (2013)
111. T.G. Myers, S.L. Mitchell, Application of the heat-balance and refined integral methods to the Korteweg-de Vries equation. Thermal Sci. **13**(2), 113–119 (2009)
112. T.G. Myers, S. L. Mitchell, Application of the combined integral method to Stefan problems. Appl. Math. Model. **35**, 4281–4294 (2011)
113. T.G. Myers, J.P.F. Charpin, S.J. Chapman, The flow and solidification of a thin fluid film on an arbitrary three-dimensional surface. Phys. Fluids **14**(8), 2788–2803 (2002)
114. T.G. Myers, J.P.F. Charpin, C.P. Thompson, Slowly accreting ice due to supercooled water impacting on a cold surface. Phys. Fluids **14**(1), 240–256 (2002)
115. T.G. Myers, S.L. Mitchell, G. Muchatibaya, Unsteady contact melting of a rectangular cross-section material on a flat plate. Phys. Fluids **20**(10), 103101 (2008)
116. T.G. Myers, S.L. Mitchell, F. Font, Energy conservation in the one-phase supercooled Stefan problem. Int. Commun. Heat Mass Transf. **39**, 1522–1525 (2012)
117. T.G. Myers, H.R. Ribera, V. Cregan, Does mathematics contribute to the nanofluid debate? Int. J. Heat Mass Transf. **111**, 279–288 (2017)
118. T.G. Myers, F. Font, M.G. Hennessy, Mathematical modelling of carbon capture in a packed column by adsorption. Appl. Energy **278**, 115565 (2020)
119. T.G. Myers, M.G. Hennessy, M. Calvo-Schwarzwälder, The Stefan problem with variable thermophysical properties and phase change temperature. Int. J. Heat Mass Transf. **149**, 118975 (2020)
120. T.G. Myers, A. Cabrera-Codony, A. Valverde, On the development of a consistent mathematical model for adsorption in a packed column (and why standard models fail). Int. J. Heat Mass Transf. **202**, 123660 (2023)

121. T.G. Myers, M. Calvo-Schwarzwalder, F. Font, A. Valverde, Modelling large mass removal in adsorption columns. Int. Commun. Heat Mass Transf. **163**, 108652 (2025)
122. H. Ockendon, J.R. Ockendon, *Viscous Flow* (Cambridge University Press, Cambridge, 1995)
123. J.R. Ockendon, A.A. Lacey, A. Movchan, S.D. Howison, *Applied Partial Differential Equations* (Oxford University Press, Oxford, 1999)
124. A. Oron, S.H. Davis, S.G. Bankoff, Long-scale evolution of thin liquid films. Rev. Modern Phys. **69**, 931–980 (1997)
125. X. Peng, J. Wickham, A. Alivisatos, Kinetics of II-VI and III-V colloidal semiconductor nanocrystal growth: 'focusing' of size distributions. J. Am. Chem. Soc. **120**(21), 5343–5344 (1998)
126. T. Peters, J. Shelton, H. Tang, P.H. Trinh, An enthalpy-based model for the physics of ice-crystal icing. J. Fluid Mech. **1001**, A12 (2024)
127. T. Peters, J. Shelton, H. Tang, P.H. Trinh, A new implicit formulation of the enthalpy method using flag updates. Int. J. Heat Mass Transf. **249**, 127166 (2025)
128. O. Petrov, I. Furó, Curvature-dependent metastability of the solid phase and the freezing-melting hysteresis in pores. Phys. Rev. E **73**, 011608 (2006)
129. G. Poots, *Ice and Snow Accretion on Structures*, 1st edn. (Research Studies Press, Chichester, 1996)
130. S.A. Prakash, R. Arivazhagan, C. Hariharan, R. Sheeja, V.A.A. Raj, R. Velraj, Review on numerical algorithms for melting and solidification studies and their implementation in general purpose computational fluid dynamic software. J. Energy Storage **36**, 102341 (2021)
131. T.Q. Qiu, C.L. Tien, Heat transfer mechanisms during short-pulse laser heating of metals. J. Heat Transf. **115**(4), 835–841 (1993)
132. H. Ribera, T.G. Myers, A mathematical model for nanoparticle melting with size-dependent latent heat and melt temperature. Microfluid. Nanofluidics **20**, 147 (2016)
133. H. Ribera, T.G. Myers, M.M. MacDevette, Optimising the heat balance integral method in spherical and cylindrical Stefan problems. Appl. Math. Comput. **354**, 216–231 (2019)
134. L.I. Rubinstein, *The Stefan Problem*, 1st edn. (American Mathematical Society, Providence, 1971)
135. J.S. Russell, *Report on Waves*. Fourteenth meeting of the British Association for the Advancement of Science. Held at York, September 1844 (1845), pp. 311–390
136. N. Sadoun, E.-K. Si-Ahmed, P. Colinet, On the refined integral method for the one-phase Stefan problem with time-dependent boundary conditions. Appl. Math. Model. **30**(6), 531–544 (2006)
137. K. Schulte, B. Weigand, On the analytical modelling of the initial ice growth in a supercooled liquid droplet. Int. J. Heat Mass Transf. **127**, 1070–1081 (2018)
138. A. Sellitto, F.X. Alvarez, D. Jou, Temperature dependence of boundary conditions in phonon hydrodynamics of smooth and rough nanowires. J. Appl. Phys. **107**(11), 114312 (2010)
139. H.W. Sheng, G. Ren, L.M. Peng, Z.Q. Hu, K. Lu, Superheating and melting-point depression of Pb nanoparticles embedded in Al matrices. Phiolos. Mag. Lett. **73**(4), 179–186 (1996)
140. M.E. Siemens et al., Quasi-ballistic thermal transport from nanoscale interfaces observed using ultrafast coherent soft x-ray beams. Nat. Mat. **9**(1), 26–30 (2010)
141. S.H. Simon, *The Oxford Solid State Basics*, 1st edn. (Oxford University Press, Oxford, 2013)
142. V.P. Sobolev, P. Schuurmans, G. Benamati, Thermodynamic properties and equation of state of liquid lead and lead-bismuth eutectic. J. Nuclear Mat. **376**(3), 358–362 (2008)
143. A.D. Solomon, Mushy zones, in *Encyclopedia of Physical Science and Technology*, ed. by R.A. Meyers, 3rd edn. Chapter IIIa (Academic, Cambridge, 2001)
144. A.D. Solomon, D.G. Wilson, V. Alexiades, A mushy zone model with an exact solution. Lett. Heat Mass Trans. **9**, 319–324 (1982)
145. J. Stefan, Über die diffusion von säuren und basen gegen einander. Sitzungsberichte der Mathematisch-Naturwissenschaftlichen Klasse der Kaiserlichen Akademie der Wissenschaften in Wien. Mathem.-naturw. **98**(2A), 616–636 (1889)

146. J. Stefan, Über die theorie der eisbildung, insbesondere über die eisbildung im polarmeere. Sitzungsberichte der Mathematisch-Naturwissenschaftlichen Klasse der Kaiserlichen Akademie der Wissenschaften in Wien. Mathem.-naturw. **98**(2A), 965–983 (1889)
147. J. Stefan, Über die verdampfung und die auflösung als vorgänge der diffusion. Sitzungsberichte der Mathematisch-Naturwissenschaftlichen Klasse der Kaiserlichen Akademie der Wissenschaften in Wien. Mathem.-naturw. **98**(2A), 1418–1442 (1889)
148. J. Stefan, Über einige probleme der theorie der wärmeleitung. Sitzungsberichte der Mathematisch-Naturwissenschaftlichen Klasse der Kaiserlichen Akademie der Wissenschaften in Wien. Mathem.-naturw. **98**(2A), 473–484 (1889)
149. H. Su, J.D. Dixon, A.Y. Wang, J. Low, J. Xu, J. Wang, Study on growth kinetics of CdSe nanocrystals with a new model. Nanoscale Res. Lett. **5**(5), 823 (2010)
150. V. Voller, M. Cross, Accurate solutions of moving boundary problems using the enthalpy method. Int. J. Heat Mass Transf. **24**(3), 545–556 (1981)
151. V.R. Voller, M. Cross, N.C. Markatos, An enthalpy method for convection/diffusion phase change. Int. J. Numer. Methods Eng. **24**(1), 271–284 (1987)
152. B. Weigand, J. Braun, S.O. Neumann, K.J. Rinck, Freezing in forced convection flows inside ducts: a review. Heat Mass Transf. **32**, 341–351 (1997)
153. B. Wetton, Enthalpy methods for moving boundary problems, in *British Applied Mathematics Colloquium* (2023). https://personal.math.ubc.ca//~wetton/talks/BAMC.pdf
154. F.M. White, *Viscous Fluid Flow*, 1st edn. (McGraw-Hill, New York, 1991)
155. Wikipedia, January 1998 North American ice storm. https://en.wikipedia.org/wiki/January_1998_North_American_ice_storm. Last accessed: 18 Jun 2025
156. A.S. Wood, A new look at the heat balance integral method. Appl. Math. Model. **25**, 815–824 (2001)
157. B. Wu, S.W. McCue, P. Tillman, J.M. Hill, Single phase limit for melting nanoparticles. Appl. Math. Model. **33**, 2349–2367 (2009)
158. B. Wu, P. Tillman, S.W. McCue, J.M. Hill, Nanoparticle melting as a Stefan moving boundary problem. J. Nanosci. Nanotechnol. **9**(2), 885–888 (2009)
159. H. Yoo, Analytical solutions to the unsteady close-contact melting. Int. J. Heat Fluid Flow **43**, 1457–1467 (2000)
160. M. Zeneli, A. Nikolopoulos, S. Karellas, N. Nikolopoulos, *Ultra-High Temperature Thermal Energy Storage, Transfer and Conversion*, Chap. 7: Numerical methods for solid-liquid phase-change problems (Woodhead Publishing, Sawston, 2021), pp. 165–199
161. J. Zhong, L.H. Zhang, Z.H. Jin, M.L. Sui, K. Lu, Superheating of Ag nanoparticles embedded in Ni matrix. Acta Materiala **49**(15), 2897–2904 (2001)

Index

T. G. Myers, *Theoretical and Practical Stefan Problems*, Interdisciplinary Applied Mathematics 63, https://doi.org/10.1007/978-3-032-04826-4

MIX
Papier aus verantwortungsvollen Quellen
Paper from responsible sources
FSC® C105338

If you have any concerns about our products,
you can contact us on
ProductSafety@springernature.com

In case Publisher is established outside the EU,
the EU authorized representative is:
Springer Nature Customer Service Center GmbH
Europaplatz 3, 69115 Heidelberg, Germany

Printed by Libri Plureos GmbH
in Hamburg, Germany